1988

ICMI Study Series Editors A.G. Howson and J.-P. Kahane

Mathematics as a Service Subject

Edited by
A. G. Howson
J. - P. Kahane
P. Lauginie
E. de Turckheim

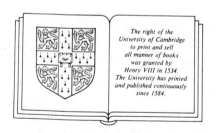

The right of the University of Cambridge to print and sell all manner of books was granted by Henry VIII in 1534. The University has printed and published continuously since 1584.

CAMBRIDGE UNIVERSITY PRESS
Cambridge
New York   New Rochelle
Melbourne   Sydney

Published by the Press Syndicate of the University of Cambridge
The Pitt Building, Trumpington Street, Cambridge CB2 1RP
32 East 57th Street, New York, NY 10022, USA
10 Stanford Road, Oakleigh, Melbourne 3166, Australia

© Cambridge University Press 1988

First published 1988

Printed in Great Britain at the University Press, Cambridge

British Library cataloging in publication data:

Mathematics as a service subject.
1. Applied mathematics
I. Howson, A.G. (Albert Geoffrey), *1931-*
510

Library of Congress cataloging in publication data: available

ISBN 0-521-35395-5 Hardcover
ISBN 0-521-35703-9 Paperback

CONTENTS

Foreword		IV
The Editors	On the teaching of mathematics as a service subject	1
J.-M. Bony	What mathematics should be taught to students in physical sciences, engineering, ... ?	20
H.O. Pollak	Mathematics as a service subject – Why?	28
F. Simons	Teaching first-year students	35
R.R. Clements	Teaching mathematics to engineering students utilising innovative teaching methods	45
J.H. van Lint	Discrete mathematics: some personal thoughts	58
H. Murakami	Mathematical education for engineering students	63
E. Roubine	Some reflections about the teaching of mathematics in engineering schools	70
M.J. Siegel	Teaching mathematics as a service subject	75
A Final Statement		90
List of Participants		91
Contents of Selected Papers on the Teaching of Mathematics as a Service Subject		92

FOREWORD

The present study, the third of the ICMI series, is the result of a cooperation between the Committee on the Teaching of Science of the International Council of Scientific Unions (ICSU-CTS) and the International Commission on Mathematical Instruction (ICMI). It is based on the work of a Symposium held in Udine (Italy), from 6 to 10 April, 1987, at the International Centre for Mechanical Sciences (Centre International des Sciences Mécaniques = CISM).

The study began by a careful investigation about the way mathematics is taught to students in another major subject in a few typical universities: Eindhoven Technical University in the Netherlands, Jadavpur University of Calcutta, India, Eötvös Lorand University and several other institutions in Budapest, Hungary, Florida Agricultural and Mechanical University in the USA, University College, Cardiff, U.K., University of Southampton, U.K., Université de Paris-Sud à Orsay, France. The past and current presidents of ICSU-CTS (the physicist Charles Taylor and the biologist Peter Kelly) took part in the program committee, which included also the president and the secretary of ICMI, the mathematicians Tibor Nemetz and Fred Simons, the statistician Elisabeth de Turckheim, and the physicist Pierre Lauginie. The Program committee issued a discussion document, which was circulated to all national representatives of ICMI, and to various institutions. It was published in the journal L'Enseignement Mathématique, tome 31 (1986), pp. 159-172, and it also appeared in French, Italian and Spanish versions. Abstracts or quotations appeared in other scientific or vocational journals, it was discussed among members of several scientific institutions (including the Académie des Sciences de Paris) and among professionals, for example the Fondation Bernard Grégory. Contributions to the discussion were received from many countries; some are reprinted in this text, others, including the discussion document, in the volume of Selected Papers to be published by Springer Verlag (see p. 92).

The meeting in Udine was attended by 37 participants, on invitations issued by the program committee. The generous hospitality of CISM – located in a beautiful historical mansion – and the working atmosphere made this symposium pleasant and profitable. The main reports – by Bony, Murakami, Pollak, Simons – are in the present book, and are preceded by a survey article written by the four editors of this volume.

Financial help was received from UNESCO, ICSU, IMU (International Mathematical Union), CISM, the Royal Society, the Ministére de l'Education Nationale of France, IBM-Europ, IBM-France, and many universities or institutions which contributed to the expenses of participants. We sincerely thank all of them, and we hope that the success of this study will prove, once again, that international actions of this type meet a real need and have an important effect.

November 1987

A.G. Howson
J-P. Kahane

ON THE TEACHING OF MATHEMATICS AS A SERVICE SUBJECT

> La science est continuellement mouvante dans
> son bienfait. Tout remue en elle, tout change,
> tout fait peau neuve. Tout nie tout, tout
> détruit tout, tout crée tout, tout remplace
> tout. La colossale machine science ne se
> repose jamais; elle n'est jamais satisfaite.
> .... Cette agitation est superbe. La science
> est inquiète autour de l'homme; elle a ses
> raisons. La science fait dans le progrès le
> rôle d'utilité. Vénérons cette servante
> magnifique.
>
> Victor Hugo, L'art et la science, in William
> Shakespeare (1864).

The title of the study may shock. Mathematics is the most ancient of the sciences. Why should it be in the service of others, or worse still, in the service of technical activities? In reducing mathematics to a service rôle, does one not belittle its contents, its image? Let us immediately state that in our view 'mathematics as a service subject' does not imply some inferior form of mathematics or mathematics limited to particular fields. We mean mathematics in its entirety, as a living science, able - as history has ceaselessly shown - to be utilised in, and to stimulate unforeseen applications in very varied domains. The teaching of mathematics to students of other disciplines must now be accepted as a fact, a social need and, also, a relatively new problematic issue. In this introduction we shall try to show the extent of the phenomenon, the social needs which it expresses, foreseeable developments and likely results in terms of choice of subject matter and teaching methods, and finally the size of the new problems which confront students and teachers alike.

I Everywhere there is a need for mathematics

1 As is common knowledge, the chance fact that there was personal enmity between Nobel and Mittag-Leffler resulted in there being no Nobel prize for mathematics. But there are still Nobel prize-winners who are mathematicians. The Nobel prize for chemistry was awarded in 1985 to two mathematicians, H.H. Hauptman and J. Karle, for the development of methods for the determination of crystalline structures, based on Fourier analysis and probability. To quote W. Lipscomb, who presented the prize: "The Nobel prize for chemistry is all about changing the field of chemistry. And this work changed the field." Before that, G. Debreu and L.V. Kantorovič had been awarded the Nobel prize for economics for work which was also of a mathematical nature. Mathematics cannot be excluded from the family of sciences. It is an integral part of scientific thought, a necessary component of contemporary advances in all scientific fields.

In physics the link with mathematics extends to Galileo and before: mechanics, optics, electro-magnetism, relativity, quantum theory are inseparable from the calculus, the geometry of surfaces, partial differential equations, non-euclidean geometries, Hilbert spaces, ... The use of computers has recently given physics an intermediate component between theory and experience, simulation, which is very close to a purely mathematical activity, experimentation on a computer. Simulation in physics and mathematical experimentation often operate on the same objects in the same way. They create new subjects of interest and cause old ones to be resurrected, for example, fractal geometry.

Informatics has a profound influence on all sciences. Yet the link with mathematics is essential. The first computers were the realisation of Turing's ideal machine. Mathematical problems are a testing ground for informatics. The algebraic and algorithmic aspects of mathematical theories are benefitting as a result. Discrete mathematics takes on a new significance. Parallel processing already suggests research avenues involving combinatorics and differential geometry.

Chemistry is a source of many difficult mathematical problems - as Hauptman and Karle's success shows.

Medicine uses sophisticated tools which necessitate cooperation between physicians, physicists and engineers. Mathematics can form a common reference point.

Biology, like economics, makes use of statistical models. Linguistics, geography and geology all use concepts and techniques which demand a solid grounding in mathematics for true mastery.

Engineers, whatever their special branch of activity, have to calculate, to construct models, to test hypotheses. Many technical problems, ranging from coding (for bank or military purposes) to geological prospecting, lead to, and draw upon, important research areas in 'pure' mathematics: the divisibility of integers, the search for prime factors, the theory of 'wavelets' in signal analysis.

        2   That is not all. Mathematical concepts now surface at all levels of social life. Let us take three simple examples.

Each individual is now confronted with an avalanche of numerical data and constant changes of scale (from the price of goods and other purchases to the National Budget, from atoms to galaxies, from nanoseconds to geological time spans). The conceptual tools needed to master such data and changes of scale exist: they are, at a fundamental level, numbers written in standard (exponential) notation and, at a higher level, geometric representations and data analysis. Changes of scale in the exploration of figures - a kind of intellectual zoom - correspond to modern notions of measure and dimension. To understand our different environments, modern ways of 'calculating', and modern views of geometry and analysis offer remarkably well-adapted tools.

For the individual, for groups, and for humanity as a whole, the evaluation of risks (car accidents, nuclear accidents, geological catastrophes) has become a necessity and it alone enables us to make rational decisions. Probability is the means appropriate to such evaluations. The concept of probability allows us critically to examine data and suppositions. Acquiring an understanding of probability should be – and how far we are from attaining this goal – a key element in the development of general critical faculties.

In the fields of production and the service industries, information technology and automation have caused programming and 'control' to become essential activities. The conceptual tool adapted to programming is the algorithm, that is to say, a systematic procedure enabling us to solve a whole class of problems. Thus, an algorithm is a means of governing thought which is well adapted to the governing of machines.

Arithmetic and geometry, analysis, probability, algebra and in particular algorithmics have a totally different meaning today, and offer far more, than they did two centuries ago – a time when discussion still occurred on whether $\overline{}2$ could really be called a number. Some modern concepts must now become part of common consciousness, and they must be integrated into higher and professional education if space cannot yet be found for them in the general school curriculum.

   3   In higher education mathematics is now taught to a wide range of students – diverse on account of their backgrounds and also because of their specialisms and aspirations. At the beginning of the century one could, as ICMI did in 1911, confine attention to the teaching of mathematics to engineers. Today, however, one must take into account the needs of all future professionals: architects, doctors, managers, etc. Whatever the major subject studied, mathematics has become a necessary adjunct: that is what we call 'mathematics as a service subject'.

The topic is important – in respect of a general concept of mathematics education – for at least three reasons.

Numerically, it involves many mathematicians in higher education (in some institutions in Canada up to 80% of mathematics teaching is to students studying other disciplines).

Socially, it corresponds to the impact of mathematics on all aspects of everyday life.

Intellectually, it forces us to look at things from a new angle – for instance, to perceive that there are many routes by which one can come to mathematics.

Nevertheless, the variety of situations is extremely large. Not only because of the diversity of students already remarked upon, but also because of different national traditions and variations in the structure of institutions of higher education. A general description becomes almost impossible. The papers which follow, and more particularly those in the volume of Selected Papers, will give the reader some idea of this wide variety.

Although important and varied, the field is relatively ill-known. In general, professional mathematicians do not see the service-teaching in which they participate as particularly rewarding. It is a part of their activity which is often hidden, in particular from students of mathematics and from future teachers of mathematics. As a result, secondary school teaching does not benefit from innovations introduced into universities as part of service teaching.

4 Let us stress once again the social importance of this service teaching - its importance goes well beyond meeting an explicit social demand. Nowadays this social demand is expressed mainly through the voices of employers and through those of colleagues in other disciplines.

Some indications of the demands of employers is to be found in the paper provided by the Bernard Grégory Association (Selected Papers). We note one surprising fact: when engineers from the French Electricity Board were asked which discipline they felt nearer to, mathematics or physics, 90% chose the former and only 10% the latter. The paper by Henry Pollak in this volume illustrates the enormous need for mathematical training which arose in Bell Laboratories, and the magnitude of the present demand.

Demands from sundry disciplines are clearly very varied and depend not only on the disciplines studied but also on the level at which studies are taking place. One can distinguish two types of subject. In those disciplines of the first type - physics, astronomy, theoretical chemistry, parts of engineering science - certain essential concepts are mathematical in nature, data are treated in a quantitative way, numerical solutions are obtainable to given problems through the use of mathematics: it can be said that mathematics permeates the whole of the discipline. Within the disciplines of the second type - biology, economics, etc. - mathematics sheds light on certain concepts and is used to set up or exploit quantitative models, often far removed from reality. The attitudes of students towards mathematics tends to differ greatly according to the type of their major discipline. This gives rise to equally different pedagogical problems.

5 To sum up:

(a) More than ever, and increasingly so, mathematics interacts with other sciences and with technical activities in which science is strongly represented.

(b) A part - changing - of mathematics represents an integral part of the general culture of each age. In our time no individual should be deprived of this component.

(c) Mathematics as a service subject represents a very important activity within institutions of higher education, a very varied, very interesting and ill-understood activity.

(d) Explicit demands for mathematics to be taught as a service subject are already important and they are growing. According to career aspirations and choice of major discipline, mathematics appears sometimes as indispensable, sometimes as useful but of secondary importance. Ways of teaching must be adapted to match these different types of demand.

II <u>What is changing, what is to be done, and why</u>?
1 <u>We live in a rapidly changing world</u>, the sciences are advancing, technologies are changing, societies are being modified. The result is new problems, new means and possibilities, and new needs. Let us quickly review what implications this has for mathematics.

As in other sciences, the output in mathematics (as measured by published papers) is increasing exponentially: it has regularly doubled every decade since the beginning of this century. In 1987 we are talking of more than 100,000 papers. It may very well be that this trend will continue, but in other forms (for otherwise where will all the paper come from!) as an increasing number of countries get involved in mathematics research.

Is the social assimilation of new knowledge on this scale possible? That is a big question which can also be posed for other sciences. Yet true scientific progress is the concurrent development of knowledge, its dissemination, and its assimilation by the public at large. The question then is to pass from development to progress. It concerns society as a whole and each individual in particular. It is not possible for everyone to know everything; but we should not believe that even today's specialised knowledge will remain permanently out of reach of most people.

Information technology has come into being at a most opportune time, for through it new ways of storing, processing and disseminating data have become possible. We therefore have new means of conserving and communicating acquired knowledge. Yet <u>consultation</u> of documents stored on disks and cassettes, and the <u>diffusion</u> of new knowledge necessitate improved means of catologuing and <u>listing</u>, and the production of abstracts, syntheses, manuals and guides. These intellectual tools, which emerge slowly but surely, are the indispensable accompaniment of purely technical tools. They will quite likely continue to appear in printed form; they will be used in research, in production and in teaching. Contrary, perhaps, to a generally accepted notion, new technologies are not going to make recourse to books redundant in the process of teaching. They will make such recourse more necessary than ever.

On the other hand, new technologies give rise both to new possibilities
and new demands for mathematical research and for the teaching of mathematics. The first volume in the ICMI study series dealt with the
influence of computers and informatics on mathematics and its teaching.
Let us extract from that a few ideas which directly concern the teaching of mathematics as a service subject.

New possibilities: the use of the computer allows one to illustrate
concepts and methods (for example, differential equations), to experiment (an essential phase of research), to assist memory and replace
technical virtuosity (in particular, with the rise of symbolic manipulation), to adapt learning to the potential and rhythm of an individual
student (CAL).

New demands: in their professional life, computer-users must know what
to ask of computers and how to interpret the results they obtain.
Those users must, therefore, have at their fingertips knowledge and
concepts which are more varied than hitherto. Computers can free the
users from most mechanical drudgery related to the learning of mathematics (memorisation, execution of algorithms), but they demand more
imagination, creativity, critical faculties (conception of algorithm,
stability, sensitivity to initial conditions, detection of errors,
control and exploitation of results). In particular, statistical software is becoming a more and more familiar object and the demand for an
understanding of statistical methods is becoming more explicit.

Fortunately, when confronted with these new demands, mathematics has
produced and is continuing to produce a flow of general concepts and
powerful methods. The development of science is not merely an accumulation of knowledge but a permanent restructuring. It is this restructuring which enables mathematicians not to get lost in the mass of
contemporary output, and enables students to assimilate a non-neglig-
ible part of mathematics rapidly and deeply. To put it more precisely,
it is when general concepts and powerful methods are brought into
being - which are the big intellectual tools for the world of machines
and the world of men - that the problem of their assimilation begins.
Choices become necessary in the subjects to be taught and new methods
of teaching have to be introduced.

        2 Let us examine how the choice of subjects presents itself.
It obviously depends upon the future profession of the students and on
the teaching they receive in their major disciplines.

There are two possible criteria. The first is to choose the subjects
that one imagines will be those most useful in the course of the
students' future professional life. The second is to teach what is
immediately usable by students in their learning of their major
discipline.

The second criterion is often what colleagues teaching the major
discipline spontaneously demand: the necessary mathematics is that

which we need, and it should be supplied at a speed to match the demands of our teaching. It can also correspond to the demands of students who seek a certain coherence between the mathematics teaching and those courses which are utilising mathematics. Such demands can, in many cases, induce mathematicians to improve the choice of topics which they teach, the order of presentation and the way in which they introduce or illustrate mathematical concepts. It can provoke a requestioning of certain habits and of traditional curricula. Nevertheless, it often leads to the formulation of impossible demands (for example, the chemist or physicist may wish to use functions of several variables long before the mathematician has been able to introduce them). Above all, its essential weakness is to ignore the first criterion.

It is this first criterion which should be the fundamental one. But it means that choice must depend upon a future of which we are ignorant. It is, therefore, hazardous and it necessitates, even much more than in attempting to satisfy the second criterion, that mathematicians work in close cooperation with colleagues working in the major discipline. Very often those colleagues demand of mathematics something other than a justification of the use they make of it. They wish their students to learn mathematical modes of thought and the mathematician's various modes of expression: abstract exploration, geometric representation, an intuition into the calculus, then logical deduction and formal rigour. The reader, for example, can find in the Selected Papers volume, the views of physicists from the Paris Academy of Sciences who demand that geometry should once more be given a prominent place in mathematics for physics students, since geometric intuition is essential for the physicist. There is an equally explicit demand from the engineers (as is shown, for instance, by Pollak, Aillaud, Roubine and Sinha).

Confronted with these two criteria, the mathematician can legitimately take the initiative. Very often, what one can and must teach nowadays depends upon discoveries or formulations made in the last thirty years, and, therefore, unknown when many colleagues in other departments were students. The mathematicians are in a position, therefore, where it is up to them to formulate proposals.

3 Let us begin by proposing an exclusion from existing courses, that of most exercises on differentiation and integration, partial fractions, inverse trigonometric functions, etc. These even today represent a large part of many service courses for first year students. They do not aid the learning of analysis (in which we include ODEs, PDEs, numerical and harmonic analysis); indeed, they often obscure it. The good way of performing this type of calculus is as a branch of algebra and it is desirable that students should be more conversant with underlying principles in order that they should better understand the origins and the use of software for symbolic manipulation which is now replacing calculations done by hand. It can also be argued that the repetition of such 'computing by hand'

when the answers are readily available on a machine is simply a waste
of time.

One single example, well discussed and analysed, can be more instructional than a host of repetitive exercises. The time saved could be
used to familiarise students with notions fundamental to analysis (for
instance, vector fields and line integrals should be substituted for
repetitive exercises on first order differential equations) and/or for
learning some discrete mathematics.

The study on the influence of computers (ICMI Study 1) highlighted the
rapid development of discrete mathematics and proposed its introduction
into the curriculum. Recommendations to this effect have also been
made by an ad hoc committee of the American Mathematical Society (see
the paper by Martha Siegel). Jack van Lint succinctly presents in this
volume a stimulating vision of what discrete mathematics can mean and
how it can contribute to the solution of problems with very varied
origins. It is a mathematical field which has never had more than a
foothold in the curriculum yet van Lint's examples show how essential
such knowledge now is for engineers. It offers a new and interesting
way in which to approach certain algebraic topics and aspects of the
theory of numbers (in particular, permutation groups and finite
fields).

The introduction of discrete mathematics may seem quite alien to the
desire expressed by physicists, which we mentioned earlier, to see
greater emphasis given to geometry. In fact, geometry - if we understand the term in its broad sense (see, for example, G. Chatelet's
contribution to the Selected Papers) - applies to the discrete as well
as the continuous. Its importance in physics, and in many other human
activities, proceeds, classically, from what physicists call symmetries
and mathematicians see as invariants under a group of transformations.
According to Chatelet - who makes reference to very remarkable texts by
Hamilton and Maxwell - fundamental geometric concepts express actions
rather than visions. Thus, vectors, arrows, diagrams express actions
as also do fibrations and parallel transports. The significance of
geometric intuition is that it represents thought in action. Whatever
the choice of geometric concepts to be taught, and this cannot be the
same in physics, engineering and architecture, the active aspect of
geometrical thought must be preserved.

So far as the choice of subjects to teach to physicists is concerned, in
particular analysis, the meeting clearly showed that merely enumerating
desirable topics leads nowhere and that, on the other hand, it is
possible to hold a constructive debate around fundamental questions:
disparate subjects or unifying concepts, ad hoc procedures or powerful
methods, fidelity to tradition or a modern approach. J.L. Bony openly
pleads for unifying concepts and powerful modern methods. The examples
he quotes are excellent, but they are only examples. The interest in
this approach is not to determine a particular choice, but to establish
a method by which one can choose.

Without pretending to cover the whole spectrum of mathematics, we must, nevertheless, make a special mention of probability. Probability comes in, or should do so, at all four levels of mathematical need identified by Pollak: everyday life, intelligent citizenship, professional work and general culture. It is not reasonable that a student should leave university - whatever his field of studies - without ever having learned of probability.

    4 It is a striking fact that non-mathematicians - even more than most mathematicians - insist on the power and value of a mathematical mode of thought. The idea is expressed equally forcefully by biologists ("never mind what you teach: teach students to reason well") and by engineers (see, for example, Aillaud, Pollak, Roubine). Let us mention, however, the reservation expressed by Tonnelat: 'Mathematical thinking is a good servant, but a bad master'. The ways of thinking acquired in the course of studies will, however, serve strictly to determine an individual's ability to update his/her knowledge in the years of professional life. By this we mean a kind of continuous retraining. Let us borrow an example from G. Aillaud: an engineer trained in combinatorial arguments will easily adapt to operations research, programming, expert systems, but he would be totally blocked should he wish to move from combinatorics to numerical analysis.

The consequence of this is that in the choice of subjects one must think not only of the knowledge we wish our students to acquire, but also of the modes of thought associated with those topics.

    5 Again it is the experience of engineering departments which particularly attracts our attention to the other side of the coin (cf. Pollak, Siegel, Aillaud): the importance of knowledge itself, as distinct from the ability to make use of it. In the course of his professional life an engineer will rarely have to solve a mathematical problem, but he will frequently have to recognise whether a question confronting him is capable, or not, of being modelled, of being treated mathematically. As in any other science, the important thing for him is to know enough mathematics to be able to consult a mathematician and to derive the most benefit from this.

The consequence of this is that in the choice of subjects to be taught one must think not only of mathematical modes of thought, but of the large range of knowledge required to permit a professional to know what might be mathematically tractable.

    6 Each professional activity demands a particular type of mathematical culture (mathematical literacy) which enables one to be an intelligent user of mathematics. This means an ability (i) to read the mathematics used in the literature of one's profession, (ii) to express oneself using mathematical concepts, (iii) to consult references or competent mathematicians should the need arise. In biology and the human sciences, for example, a need frequently ex-

perienced is to be able to use mathematics as a language to express the problems of the discipline. This concept of a mathematical culture or a type of familiarity with mathematics peculiar to each discipline or each profession seems to us better suited to present needs than that, frequently used, of a knowledge of a 'fundamental' range of techniques. Indeed, this knowledge of a range of basic techniques must be modified as mathematical culture is acquired: they are only fundamental with respect to a particular goal, and this end seems to us to be the mathematical culture in itself, varied and variable in the same way as activities and technologies.

Mathematical culture must unite these two distinct aspects: mathematical modes of thought and a range of essential knowledge.

   7 We have chosen to insist on what is changing and what, as a consequence, forces us to modify curricula. In the same way that societies, technologies, sciences, mathematics are not going to stop changing, it would appear that, in the future, curricula will constantly have to be modified. Will this upheaval in curriculum design result in the developing countries being left permanently behind? This anxiety was expressed at the Udine meeting and it must be given serious consideration.

Even, and above all, in developing countries sclerosis of the curriculum will prove a catastrophe. Everywhere, then, we must be on the alert to track down what is - and what needs - changing. But that is not to say that teaching programmes must change everywhere in the same way. It would be absurd to attempt to teach new topics if there is no one fitted to teach them. The choice of topics, at university level, must be made by the teaching staff bearing in mind their levels of competence, their fields of interest, and other circumstances peculiar to their actual situation.

   8 Let us end with the underlying idea which has provided cohesion to this study. Much more than in the past, and more and more so, thanks to the influence of computers, users need to understand mathematics, to assimilate concepts rather than techniques. Let us stress that this demand is expressed with particular force by engineers who are especially sensitive to the effects of rapid technological change. The rôle of teaching is to prepare students for change and, on the whole, they are ready to recognise this aim as essential. Such preparation for change is necessary whatever the student's future professional activity will be. Thus there is no contradiction - indeed quite the reverse - between a teaching devoted more to fundamentals and a teaching nearer to practice.

More fundamental, more practical, less technical; it seems to us that these trends should obtain as a general rule for the teaching of mathematics as a service subject.

   III What is being done and could be done. With whom? How?
    1 We have just written of technologies, of sciences, of

subjects and of curricula. Yet at the meeting in Udine, the main part
of the discussion centred on another aspect: teaching and learning
methods, pedagogical experiences and problems, the relationships between teacher and students, and the social function of those engaged
in service teaching.

We shall consider in order:

- students, from those of the first year to those in continuous
  education (2,3,4),

- desirable directions for developments (concerning, inter alia,
  mathematical reasoning, rigour, theory and examples, and
  modelling) (4-11),

- tools (computers, books, examinations) (12,13,14),

- relationships between teaching colleagues and those within
  the mathematical community at large (15-21).

        2 The entry of students to higher education deserves special
attention and Fred Simons' paper is devoted to this topic. Let us
abstract from it a few topics which he describes.

It is a remarkable and somewhat paradoxical fact that first-year
syllabi should be practically the same throughout the world for all
service-teaching to students of engineering and the physical sciences.
Yet students come to university with very different levels of attainment. Some of them, ill-trained during their secondary schooling, find
themselves in difficulty on courses which their peers find accessible.
Two solutions have been mentioned: imposing more strictly a minimum
level of attainment on entry (a move which would often run against
national traditions and mores) and organising special entry programmes.
These last, 'booster', courses have given rise to some interesting
experiments, but there appears to have been little done in the way of
evaluation. In any case an essential effort is required to spell out
the prerequisites to first-year teaching by giving precise indications
on what subjects will be needed, when they will be used, and in what
context. In some places, such clarification of prerequisites, together
with the production of complementary documentation and the establish-
ment of booster programmes has already occurred, and been welcomed by
students and colleagues (see Shannon, Selected Papers). This is also
a useful way in which to help and influence secondary schools.

An example of another experiment in first-year teaching can be found
in Southampton. This is self-paced, individualised instruction (with
opportunities for consulting tutors) which is controlled by means of
tests taken at the end of each 'unit'.

Numerically - whether in terms of the number of students involved, the
number of lecturing hours, or the number of lecturers - the teaching

of first-year service courses is of considerable significance. It is
at this level that the most crucial factors common to all service-
teaching commitments arise: student motivation and that of their
lecturers. It is at this level that an ill-adapted course can so
easily deprive students of an interest in mathematics and can conceal
from them the true flavour of, and creativity inherent in, the subject.
It is also at this stage that vocations can reveal themselves. This
is then a time when the need to exercise 'pedagogical care' (see
Martha Siegel) is uppermost. It is, clearly, a level at which considerable pedagogical research is needed. One can assume that students
of mathematics enter university motivated to study the subject (although how long that motivation will last will depend very much upon
the courses they are then given); but for those taking mathematics as
a service subject it is usually necessary to create/foster motivation.
Yet it is at this stage that lecture rooms are at their most packed -
a time when the need for small classes and tutorial groups is at its
greatest. Where sequential courses are not set out from the start, it
is also the stage at which students will have the opportunity to opt
for different career directions - and this brings a corresponding need
for multivalent types of mathematics teaching. The first-year, too,
is often the time when mathematics is used as a sieve to separate out
the 'clever' from the 'dull' students. Assessment then becomes over-
important with the result that students devote their major intellectual effort to cramming for the end-of-year examinations.

3 At the other end of the time-scale, continuous education
is now a fascinating field in which there are already many valuable
experiments to report. Yet it is still an insufficiently explored
area. The account of the development of continuous education in Bell
Laboratories is well worth studying (Pollak). Here, motivation is
clear. But the teaching approaches most suitable for adults with
considerable professional expertise will differ considerably from
what is traditional practice for university academics. Students must
be given the opportunity, and encouraged, to proceed at their own pace
(books, papers, software) and the teacher should assume (more even
than elsewhere) the rôle of expert and adviser. The provision of
materials suitable for use on continuous education programmes is an
urgent need.

4 The present position so far as motivation to study is
concerned is often described in gloomy tones:

(a) users frequently demand a fantastic quantity of techniques,
of tricks, while allowing mathematics only a ridiculously small frac-
tion of the students' time;

(b) students bother only with examinations and prefer to learn
and apply formulae rather than to develop their reasoning powers;

(c) students couldn't care less about what worried Fourier or
what prompted the development of Hilbert spaces - that will not help
them to pass the examination!

Perhaps then we should work towards a situation in which mathematics
is taught so that:

      (a)   students should later be able to learn more mathematics by
themselves;

      (b)   students can see how, where and when to apply the mathe-
matics they know.

Let us consider the implications of this so far as the introduction of
concepts, mathematical reasoning, the rôle of rigour, the relationship
between theory and examples, modelling, and styles of teaching are
concerned.

     5 The most general and important concepts (convergence,
linearity, differentiability, orthogonality) are also the most difficult to assimilate. Students must learn to recognise them in very
different situations. For instance, orthogonality – a geometrical
concept – will be encountered more often in analysis than in geometry.
At the same time, if they are to be used effectively, these concepts
should be understood in their simplest form. For example (cf. Bony),
differentiability at the level most suited to the majority of service
teaching means that there exists a good first order approximation.
Linearity is at the same time a fundamental geometric concept (linear
spaces, vectors) and the study at the first order, of all that is
differentiable (linear mappings). The notion of convergence is one
that could be illustrated and clarified by means of a computer (calculations, graphics). It is a notion often obscured by recipes and
repetitive exercises which only draw upon calculating techniques. Far
better, then, that the student should be aware of it in all its riches
(speed of convergence, various examples of convergence drawn from
analysis and probability, the lack of convergence of certain natural
series – Taylor or Fourier). But it is even more important that the
student should know thoroughly certain simple and general facts (for
example, that, in a convenient, well-defined sense all Fourier series
are convergent (cf. Bony)).

     6 Mathematical reasoning is a valuable part of learning. It
is in mathematics, and in mathematics alone, that students can encount-
er a formalised hypothetico-deductive system and come to understand the
rôle of hypotheses, deductions, refutation through counter-examples,
proofs by contradiction, ... : in brief, formal logic in action. But
the logical aspect of mathematical reasoning must not obscure the
others: geometrical intuition, a search for good geometrical representation, analogies, generalisations, the study of particular cases.
(Polya's books, inspired by teaching engineering students at Zurich
Polytechnic and also architects and chemists, are a source of most
interesting suggestions on this topic.)

     7 Rigour is necessary. "Under very general conditions, such
a conclusion can be drawn" is not a mathematical statement, because
there is no means of verifying that 'the very general conditions' are

satisfied. In mathematics, rigour of language is a guarantee of rigour of thought. It is therefore crucial that students should learn to detect and criticise incorrect formulations and develop the ability to express themselves correctly, both orally and in writing. It is also crucial that mathematical rigour should work effectively for them and that they should appreciate the difference between rigour and pedantry. In fact, the most effective rigour is often exhibited through the most simple and elegant forms of expression.

Proofs are not indispensable in service teaching. Yet they are welcome if they throw light on concepts or stimulate the students' interest.

Certain teaching modules can be organised to start from the reading of books and papers (see, for example, the paper by Clements). To train students to read texts critically, to exercise control over their own means of expression, to develop rigour and elegance, calls for attention and time in amounts which are not usually available to the teacher: seeing students individually, getting them to talk mathematics, reading and correcting their writings. Teaching 'through example' is necessary; it is not sufficient.

      8 Let us insist again on the rôle of rigour in the choice of theorems to state. Statements must, as far as possible, be rigorous and <u>simple</u>. For instance, the following statements are excellent if one places them in the correct context:

      all Fourier series are convergent (context: L^2 or distributions);

      all functions integrable on $\mathbb{R}$ tend to zero at infinity (context: distributions);

      all functions from $\mathbb{R}$ to $\mathbb{R}$ are Lebesgue measurable (context: a model of set theory excluding the axiom of choice);

      any linear functional on the space $C(\mathbb{R})$ (or $\mathcal{D}(\mathbb{R})$ or $L^2(\mathbb{R})$) is continuous (same context).

The choice of correct statements necessitates close cooperation between teachers and mathematics researchers in all areas, and permanent research into what can simplify life for students and for users of mathematics.

      9 The choice of examples/applications and the most appropriate times for giving them is a matter of considerable pedagogical delicacy. To motivate students, to make them appreciate the interest and value of a particular theory, there is nothing like good examples drawn from their major field of study.

Should one begin with the examples or with the theory? Bottom up or top down? There is no universal answer. The meeting at Udine showed that each approach could be justified. An Australian colleague produced two examples: one from operations research in which it seemed appropriate to begin with an example and then to allow the

theory progressively to emerge, a second from statistics where it
seemed more effective to teach the theory first. In the former case
the students were helped to unify and harmonise their exploration of
other examples; in the latter the students became equipped to consider
a statistics question as a real problem. The choice is likely to
depend upon the level at which the student is. At the end of a period
of study, students may well be acquainted with a number of examples and
it is good to have a varied choice available so that the theory is
demonstrated in all its power. On the other hand, at the beginning of
the period of study, examples must be very significant and fully
developed.

The combination of the two approaches can, as Murakami shows in his
paper, be an excellent pedagogical device.

10   The transfer of mathematics to other disciplines generally
necessitates a knowledge of mathematical theories and of problems to
be considered, and an ability to construct models, allowing one to
transform problems posed in the major discipline into ones which can
be dealt with by the mathematician. The teaching of modelling is,
therefore, inseparable from the consideration of significant examples.
It can begin, in an elementary way, from the first years at university.
It is, however, in the course of the last years of study - if mathematicians still have a part to play at that stage - that advanced
modelling can exert its greatest influence and be of most interest:
the treatment of real problems arising in laboratories, liaising with
the world of production and services - both industrial and non-
industrial.

11   Whatever the subjects taught, one must endeavour to give
students a feeling for the beauty of mathematics at the same time as
one is demonstrating its usefulness. That feeling may be aroused by
a detail in a lecture, a well-chosen problem, an elegant proof, or a
neat enunciation of a result. It can be stimulated if the students
see the experimental and heuristical character of discovery in mathematics. It can also develop if students see the whole spectrum of
mathematics as a living science. Brief mentions of its history -
ancient and contemporary - of current developments, of the links with
philosophy and music as well as those with physics or information
technology, may, if put forward at the appropriate time, stimulate the
interest of non-mathematics-specialist students.

It is good that students should appreciate for themselves the particular way in which mathematical knowledge and understanding is acquired.
In particular, it is interesting to consider the respective rôles of a
'proof' in mathematics and an experiment in physics.

12   Computers and informatics present us with new means of
teaching. They change the relationship between teacher and student,
by making the student at the same time more active, more free, more
disposed to experiment, and the teacher more indispensable as the

expert (in mathematics! - not necessarily in the handling of hardware and software), guide and counsellor. These new relationships, together with the new possibilities (graphics, self-evaluation, computer marked assignments, computer-assisted learning) and the new problems which arise (creation of software, commercialisation, cultural, social and economic consequences) are analysed more fully in the Strasbourg study.

Without going back over, what is now becoming, old ground, let us stress again the importance of particular questions which arise in connection with service teaching. These questions are addressed in several of the papers which follow. Experiments in the use of symbolic manipulation are mentioned by Clements and by Hodgson and Muller (Selected Papers). Let us also mention the need at all levels of teaching for the provision of software correlated with books and course-notes.

13 The importance of reading has been stressed. Students must become capable of reading the kind of mathematics to be found in publications relating to their major disciplines: such readings may, indeed, be used as a source of motivation for their study. In some languages (e.g. French, German, English) mathematical literature from the early eighteenth century onwards provides a considerable resource of interesting readings. This is now beginning to be exploited by authors of books on service mathematics who reprint long extracts from classical works (e.g. 20 pages of Laplace in a recent book Probabilités pour l'ingénieur).

Yet reference books are still lacking, and so are presentations intended for a public which, though informed, does not have a specialist knowledge of mathematics. (G. Aillaud pleads the case of those engineers to whom mathematicians do not supply the means for embarking on continuous self-education.)

14 Unless new approaches are sought, examination systems are likely to block the evolution of teaching and the adoption of new methods. Yet used carefully, they may present a tool for the transformation of learning and teaching. For example, it is not difficult to imagine questions needing no technical virtuosity but which test, and so encourage, assimilation of concepts and the acquisition of a certain critical faculty (such questions can sometimes have the advantage of being presentable in a multiple-choice format). Yet thinking along these lines has only just begun.

Modelling courses can be assessed in a natural (but time-consuming) way (see Clements).

15 Most of those taking part in the meeting at Udine were mathematicians. There were also engineers (Ezratty, Roubine), a physicist (Lauginie), and a biologist (Peter Kelly, the outgoing President of ICSU-CTS). Who teaches service mathematics courses? Who should teach them? These were two of the questions asked in the

preliminary discussion document. A few points can now be made towards supplying answers. Let us consider successively three particular questions: the specific rôle of the major discipline and of those teaching it, the pedagogical importance and the necessity for the recognition - in career terms - of the work of those mathematicians engaged in service teaching, and finally the rôle in this field of the mathematical community as a whole.

16   Physics is inconceivable without mathematics and there is, of necessity, a considerable amount of mathematics in physics courses. It can be envisaged - indeed it has been tried - that physicists should be entrusted with the teaching of all the mathematics needed, and that they should control the pace at which mathematical notions are introduced. In any case, physics is a mine of topics and of illustrations which can illumine mathematics courses. For example, one of the best introductions to a course on Fourier analysis is to be found in the Berkeley physics course (Waves, p.91): press the sustaining pedal on a piano, say 'oh' in the direction of the strings and listen for the response 'ohh' from the strings vibrating on your voice's wavelength. Your voice is analysed by the strings, then synthesised. The Fourier transform - and in all its contemporary guises - is no more, no less, than the theory behind this harmonic analysis and synthesis. When mathematics courses are entrusted to a mathematician, it is extremely beneficial if physicists can lead example/problem classes; there must be liaison between physicists and mathematicians to ensure that both points of view are properly represented. The mathematician's contribution (as seen by Pierre Lauginie) is to simplify what his physics colleagues might have done; by so doing, the former's rôle can be more readily appreciated and recognised (see, for example, Bony's paper).

17   Teaching engineers is not altogether dissimilar to teaching physicists. In this volume complementary points of view on this important question can be found (Murakami, Pollak, Roubine). Let us, however, stress another aspect. Engineers form a vast pedagogical resource because of their professional expertise and the wide range of problems of which they have knowledge; this goes far beyond the mere teaching of engineers.

18   Mathematics for biologists seems somewhat different. Peter Kelly chose to classify the interactions between mathematics and biology in the following way:

- integration (example, quantitative genetics): mathematics is an integral part of the biological concept and must be taught, at the same time, by the biologist.

- disjunction (example, models for population growth): mathematics may be presented in an autonomous way in an introductory course.

- instrumental relationships (example, data analysis): mathematical techniques only constitute auxiliaries to the biology course:

they must be taught in separate units at the time when they are
needed in biology.

- conceptual relationships: it may be that mathematics gives rise
  to biological concepts or to a better understanding of biology
  (probabilities for the theory of evolution, set theory for the
  construction of taxonomies); this justifies the teaching by
  mathematicians of mathematics which is likely to prove a future
  resource for biologists.

At a practical level, it is worth making a few remarks about the remarkable success of a few instances of integrated teaching. Here is one
example. At the Agronomics Institute of Paris-Grignon, a course on the
theory of surveys and data collection has been jointly taught for several years by a statistician and an economist. The economist presented
the successive steps of a sample survey, set up by professionals, on
the economy of a wine-producing region and the statistician highlighted
the general principles of the theory underlying sampling. The connection between the examples and the general theory was made clear to
students in the open, on-going and often animated debate between the
two teachers.

   19 The greater the distance between mathematics and the major
discipline, the more the mathematician must listen to his colleagues'
concerns. Team work, with representatives from the major discipline,
provides a serious guarantee and sometimes provides the only opportunity for mathematicians to contribute to the teaching. A good
example of collaboration with architects on an advanced course on
architecture was mentioned by Shannon.

   20 It can be seen, therefore, that pedagogically (always) and
scientifically (often), service teaching when it is wholly or partly
undertaken by mathematicians demands more inventiveness and effort than
does teaching to future mathematicians or teachers of mathematics.
Those involved in it should feel that they are doing something immens-
ely valuable. Many do put into it the best of themselves - as teachers
and also, as we have hinted several times, as researchers. And yet,
within universities, work in this field receives very little consideration, particularly where promotion is concerned. The Udine meeting
provided a platform for making a demand in this respect: that
university authorities and the mathematics community in general should
give greater recognition to the services rendered by those who
specifically devote themselves to service teaching.

   21 Some colleagues devote themselves entirely to this kind of
teaching. More often than not, though, service teaching is shared out
amongst the mathematicians in any institution, possibly with the exception of those who appear to be the "purest". Now if the present study
can claim to have reached any conclusion, it is that even the 'purest'
mathematics (let us say axiomatic set theory - which we have still had
cause to mention) may still be very useful within service teaching.

Too often, the image mathematicians have of 'applied mathematics' is that which prevailed in the nineteenth century: special areas in analysis (in its wider sense), with possibly a dash of statistics. Today, geometers, number theorists, algebraists all come face to face with important applications. In particular, the need to teach discrete mathematics to millions of students demands the co-operation of mathematicians specialising in algebra, algebraic geometry, number theory and combinatorics. Let us once again turn to Polya, an analyst by training. His work leading to Polya's Theorem on combinatorial enumeration arose from a problem in chemistry - listing the isomeric forms of aliphatic alcohols. Thus being involved in service teaching can prove stimulating even at the research level and no mathematician ought to stand apart from this type of teaching.

22 Let us end by recalling the example of Nobel Prizes awarded to mathematicians working in economics or chemistry. Our study is inspired by the enormous 'service' which mathematicians are able to provide to facilitate the progress of science and mankind. Yet 'service' here should not be interpreted in the sense of subordination. On the contrary, the type of service required nowadays demands intimate cooperation between mathematicians, teachers and users. Such a partnership, for which our study calls, would do much to ensure future advances in education as well as in science.

WHAT MATHEMATICS SHOULD BE TAUGHT TO STUDENTS
IN PHYSICAL SCIENCES, ENGINEERING ... ?

J.-M. Bony
Centre de Mathématiques, Ecole Polytechnique,
91 128 Palaiseau Cédex, France.

I shall try to express a few ideas, related to what could
be the beginning of an answer to that question. A first attempt is to
make a list of domains of mathematics whose teaching would be useful.
One obtains a beautiful catalogue, with the following interesting
property: showing it to anybody, you obtain simultaneously the two
following answers: a) that is far too much, b) something (very
important) is lacking.

One has to make choices, and I shall concentrate on a few related
subjects: the distinction between scattered and unifying topics, the
question of teaching modern mathematics, the choice of the level of
the teaching.

1 DOMAINS OF MATHEMATICS THAT SHOULD/COULD BE TAUGHT

In some sense, it is an easy question, if you do not choose
between should and could, and if you keep the "domains" sufficiently
vague. You can ask users what they use, what kind of mathematics they
find important or useful. The intersection of the answers gives a very
small list, while the union of the answers gives something like the
following catalogue:

a) (at an elementary level): calculus including ordinary differential
equations, linear algebra, probability, statistics, discrete mathematics, some geometry.

b) (further): complex variables, Fourier, Laplace, convolution,
Lebesgue integral, distribution theory. Partial differential equations.
Hilbert spaces. Tensor calculus. Group theory. Special functions.
Geometry. Calculus of variations. Dynamical systems, fractals, chaos.
Stochastic process. Numerical analysis. Non linear phenomena, and so
on ... (of course, computer science is outside the scope of this talk).

The name of a domain can, however, have quite different meanings: the
word "geometry" can correspond, for instance to: a) that elementary
geometry (in the 3 dimensional space) which used to be known by students
entering university, but which is not known nowadays, b) the concepts
of algebraic and differential geometry arising in elasticity, general
relativity, (string theory!), c) the concepts arising in engineering
in drawing assisted by computers (approximations of surfaces by graphs
of splines ...), etc.

It is clear that a high level course on all these subjects would be
fantastic but is completely unrealizable. The constraints of time are
extremely strong, as well as the constraints due to the level and the
motivation of students.

How to choose between these domains? For each of them, how to decide
which precise topics should be taught, and at what level? The answer
will certainly depend on the category of students and on individuals.
But perhaps it is possible to express some general ideas on this point.

2 SCATTERED TOPICS, OR UNIFIED AND UNIFYING ONES

I do not think that the answer to "what" is "a little bit
of each domain of the catalogue". Actually, a course in physics
contains always some mathematics, and there is always in the curriculum
of each student at least one course of mathematics.

I think that a precise mathematical topic, arising just in one chapter
of physics is well placed in this chapter. It is linked to its
physical significance and this would be lost without any profit if it
were transferred to a "patchwork style" course of mathematics.

For instance, it is meaningful to make a course of mathematics studying
special functions in relation to group theory and/or Hilbert spaces
and/or differential equations. (This would be a high level course!)
But I do not see the usefulness to include in a course of mathematics
the study of such special functions or of a particular family.

On the contrary, I would like to give an example of what I call a
unifying concept, the existence of which is an important justification
for the existence of a specific course of mathematics.

It is well known that a linear operator which commutes with transla-
tions and is continuous (in a very weak sense) is a convolution operator. This means that, for a physical system considered as a "black
box", transforming a function $i(t)$ (the input) into a function $o(t)$
(the output), if o depends linearly on i , and if to $i(t-T)$ (the
same input T hours later) corresponds $o(t-T)$ (the same output, T
hours later) then one has $o = k + i$. Moreover, k is the output
corresponding to the input δ (Dirac measure) and the output corres-
ponding to $e^{i\omega t}$ is $\hat{k}(\omega)e^{i\omega t}$, where $\hat{k}$ is the Fourier transform
of k .

This result, which is also valid for functions of space or of space-
time variables occurs in every branch of physics. It explains why
convolution is so universal in physics. Each time I taught it to
students, I got a strong reaction of interest: they had some knowledge
of this, scattered in different branches of physics, and mathematics
was giving a unifying explanation.

Of course it is not only this particular result, but the whole domain:
Fourier, Laplace .... , whose unifying character in physics is obvious.

If one has to choose what should be the content of a course in mathematics, I think that priority should be given to that kind of unifying domains.

3 MODERN OR NOT

Should one teach recent mathematics or is it sufficient to teach what was taught 50 years ago? This is, of course, a key point before one goes on to ask: who teaches?

A Life has changed

This is the main reason why our teaching should change. An important example is the modifications that are/should be induced in our teaching of calculus or linear algebra by the existence of computers.

> Numerical methods have to be taught for computing integrals, solutions of O.D.E.s, of linear systems ... .

> Some topics, traditionally taught, are becoming less important, and the time devoted to them should certainly be reduced: computations of integrals via rational fractions, determinants and their use for solving linear equations, ...

> The importance of other topics, leading to some developments of calculus for instance, is increasing: estimation of error, speed of convergence, well or badly conditioned matrices, ...

> Some concepts should be introduced, at least by examples, to show what could be or could not be asked of a computer: for instance behaviour for large time and more generally qualitative properties of solutions of O.D.E.s.

> The concept of algorithm is becoming central, and many classical proofs by induction will gain if expressed in terms of the existence of an algorithm.

> The concept of convergence, the importance of which does not decrease, will gain if it is closely related to degree of approximation, and also to stability.

Some time will be gained, some time will be used in this transformation of our teaching. Calculus (for instance) will remain calculus, but we shall have to change its teaching to retain its usefulness.

B Mathematics has changed, and provides new possibilities

This is evident, and important, for new branches of mathematics, but I think it is also important for "old" subjects.

I would like to give a (bad) personal experience. I was teaching students in physical sciences, in their 2nd year at university, on the separation of variables for solving some partial differential equations such as vibrating strings. Solutions are given as a series of particular solutions, the coefficients of which are given by those of the Fourier series of the initial data.

I discovered that it was not possible to give any problem where I could ask students to prove that the series is actually a solution. I had just taught them theorems on the derivative of a limit when the derivatives converge uniformly, and, for those well-known functions (piecewise linear, or p. quadratic, or p. exponential) whose Fourier coefficients can be computed by students, the uniform convergence of second derivatives never happens.

What I was doing is clear: I was teaching 19th century mathematics because it is the tradition to do so, instead of teaching the corresponding 20th century concepts which are simpler and more powerful. Any limit (in a very weak sense) of solutions of a linear PDE is a solution, if one introduces the concept of weak solution (i.e. solution in the sense of distribution theory, but it is not necessary to pronounce the word: the definition is quite simple). The concept of uniform convergence for functions, and the related theorems on Riemann integral, are certainly over-valued in our teaching, while the simpler concept of weak limit is usually reserved for higher studies.

Is it so serious? I think so: if you require unnecessary restrictions for results which are always true in practice, students and users will think (and they will be right!) that mathematical rigour is nothing but a constraint, and that they need only a cook book made with formulas arising in mathematical theorems.

C Efficiency, rigour, efficiency of rigour

It is clear that, in service mathematics, it is not possible to give proofs of all statements. This should not be seen as a constraint due to the time or to the level of students, but as an advantage. One has just to pay attention to the simplicity or the strength of a theorem, not to the difficulty of its proof.

Should we give only correct statements? Should we make a clear distinction between the case when we are giving a (rigorous) proof, and the case when we are suggesting by other means (examples, particular cases, physical meaning ...) that a statement is true? Should we require (simple but) correct proofs of students? I think so, but only if our teaching shows that this rigour is efficient. Some examples:

a) Physicists think that the Fourier series of any periodic function f converges. They are right, and we should give a theorem saying that, the simplest being that it is true for the weak convergence.

After, we can examine what happens more accurately (Gibbs phenomenon, ...).

b)   If   Curl(X) = 0 ,   and if the topology is O.K., then   X   is a gradient. The restriction in the statement is an important part of its efficiency. Otherwise, the magnetic field created by an electric wire would have a scalar potential, which would be catastrophic for electric engines.

c)   The theory of distributions not only provides mathematical concepts corresponding to a lot of physical ones, but it gives the simplest, the more efficient theorems, as far as linear analysis is concerned. Many statements are true without any conditions, and when an assumption is required, it is a serious one.

d)   The restrictions in Lebesgue theorems on taking limits or deriva- tives under the sign of integration are important parts of their efficiency: when the assumptions are not satisfied, it is not rare at all that the conclusion is false. In this respect, the corresponding theorems for Riemann integrals do not give a good idea of the efficiency of mathematical rigour.

e)   A provocation: who has any objection about teaching, at the most elementary level, the principles of Lebesgue integration, i.e. (admitting that any subset of   R   is measurable, which is valid in a coherent model of set theory excluding the axiom of choice) giving the following statements:

-   any non negative function has an integral $\int f(x)dx \leqslant \infty$, and the usual relations with $\leqslant$ and + are valid.

-   a function   f   is integrable if f^+ and f^- have finite integrals.

-   the Lebesgue theorems about limits are true?

After that, it is easy to have, for continuous functions, the relations with primitives and to show that Riemann sums are a (not very efficient) way of computing approximations.

The gain in efficiency would be very important, and if it is true that proving Riemann theory requires less than proving Lebesgue theory, I think that accepting Lebesgue theory requires less than accepting Riemann theory.

4   WHICH LEVEL?
Having chosen to teach a domain of mathematics, you can do this at a more or less elementary level. This choice should be, of course, balanced with the constraints of time and of the level of students. But it should also be balanced with your precise purpose: what is the gain in efficiency, in depth of understanding for, say, physics or engineering? I shall start with a well known example.

Example: differentials in physics

1st Level Colleagues in physics or chemistry usually want at the beginning of the 1st year at university a teaching of differentials. Fortunately, they are usually quite satisfied with a teaching of partial derivatives, and a statement of 1st order Taylor expansion

(*) $f(u+\Delta u, v+\Delta v) - f(u,v) = f'_u \Delta u + f'_v \Delta v + \text{remainder}.$

It can be said then that the differential notation in physics is just a notation for small variations, when remainders can be neglected.

However, there is an important point on relations between mathematics and physics: physicists consider variable physical quantities, while mathematicians consider (and define partial derivatives for) functions. For instance, from Ohm's Laws $P = VI$, $P = RI^2$, one is not allowed to deduce $\partial P/\partial I = V = 2RI$. And, even with notations used in thermodynamics, one should not try to write down $(\partial P/\partial I)_{V=Cst,R=Cst}$ from a formula like $P = V^{1/2} R^{1/2} I^{3/2}$. Writing down partial derivatives with the excellent thermodynamic notations (say $(\partial P/\partial I)_{R=Cst}$) requires 2 conditions: a) R and I can vary independently, b) P is determined by R and I .

2nd Level Mathematics have quite a good definition for the differential of a function defined on an open set of a vector (or affine) space. It is the linear function which is tangent to the given one. They can be tempted to teach that to students in physics. The result is usually not very good: students cannot see the relation with differentials as used in physics.

My next provocation will be the following: students are right. Teaching that and only that is not useful and is confusing. For instance when writing down $dP = 2RIdI + I^2dR$ is it possible to answer to the 2 following questions: a) where is the function?, b) where is the vector (or affine) space?

There is just one case where this could be useful: physical quantities defined in the space (or the space-time) can be naturally considered as functions on a vector space.   However, this space is always equipped with a Euclidean (or Minkowski) structure, and the concept of gradient (which can and should be taught at the 1st level) contains the same information.

3rd Level That level (which requires the 2nd one) is the good one. Assume that the set S of possible states of your physical system is equipped with the structure of a differential manifold (the dimension is the number of degrees of freedom). Then a variable physical quantity can be considered as a function on S . A relation $P(s) = R(s)I(s)^2$ is an equality of functions on S , and dP at some point s is a linear form on the tangent space to S at s

(the space of infinitesimal variations of the system), and so on ... .
Now the mathematical and physical meanings of differentials fit
together, and everything is well justified.

This rather long example is, I hope, instructive.

- If it is true globally that high level, elaborated, modern
  mathematics is more efficient for understanding physics,
  nothing is automatic.

- An unquestionable progress in mathematical understanding
  (the 2nd level here) may be of low interest for service
  mathematics.

- Thinking about modelling, about how concepts, notations,
  results are used outside mathematics is necessary, and
  should be an important part of the teaching.

- Such an analysis does not give a unique answer. It is
  quite reasonable to teach the 1st level, with no mathematical definition of differentials and just information
  on their physical use, and it is quite reasonable to teach
  the 3rd one (simplified as far as one can) if one has time
  for it.

- The same kind of analysis is probably useful in many parts
  of mathematics. For instance, it is certainly valuable to
  teach Kolmogorov axiomatic probability theory if it is used
  to model a serious physical situation, Brownian motion for
  instance. Is it worth it if you have just applications to
  coins, urns, cards, dice (things which are already symbolised,
  which are already abstract concepts, and which may give a
  wrong idea of modelling)? At a more elementary level, is it
  not better to teach only statistics?

I was supposed to answer the question "What should be taught", and my
conclusion will be that I have no answer. There are so many domains
of mathematics whose teaching would be useful, and there are so many
constraints. However, this should not be an excuse for perpetrating
traditional teaching.

Life, and mathematics, and intervention of mathematics in life have
changed. The teaching of more modern mathematics, of domains which
were not taught before, should be considered very seriously, not for
pleasure, but for efficiency.

There is just one thing of which I am sure. It is the absolute
necessity of a collaboration, at the highest possible level, between
mathematicians and users. I have had an excellent experience, at
Orsay University, of teaching a course of mathematics to students in
physics, while exercises and problems were done by physicists. Such

teaching demands a knowledge of recent mathematics and precisely how mathematical concepts and methods were, are and could be used in other sciences or techniques. My opinion is that discussing together, at the research level, is the key to the problem.

MATHEMATICS AS A SERVICE SUBJECT - WHY?

H. O. Pollak
40 Edgewood Road
Summit, NJ 07901 USA

My assignment is to examine the question of why we teach mathematics as a service subject. This question is of course also very much in evidence in many other papers at this meeting. It will be studied in the present paper from the point of view of this particular author, who spent a 35-year career as a mathematician in industry, both as a mathematical researcher and as leader of a group engaged in research in the mathematical sciences. We begin with some anecdotes, examples of incidents in the old Bell Telephone Laboratories before 1984, which have helped to shape my thoughts about mathematics service teaching. These examples of formative experiences will be followed by some of their implications for mathematics education, and by some broader observations about the telecommunications industry and its relations to the mathematical sciences. Finally, we shall include some thoughts on the "why" of teaching mathematics in a larger sense.

Incidents

We begin, as we said, with some incidents during my career which have helped to shape and to modulate my ideas on the purposes of mathematics as a service subject.

1 I was a member of a committee considering the promotion of a young man who served in the role of a technical associate in the area responsible for semi-conductor device development. One of the exhibits which formed the basis of the promotion request was an internal memorandum he had written on the removal of impurities in a semi-conductor material. It was an interesting piece of experimental work, but near the end of the write-up, he mistakenly divided by $10**7$ instead of multiplying by it, and so arrived at 3 impurity atoms per cc rather than $3*10**14$. Neither he nor the people above him in our chain of command noticed this totally unreasonable impurity level at the conclusion of the memorandum.

2 An engineer came to see me, and asked me if I could sum the following series for him (or one very much like it, I am not sure):

$$\sum_{n=2}^{\infty} \{1/n + [1/n**2 + 3/n(n-1)]**\frac{1}{2}\}**n.$$

Pollak: Why?

According to my experience, this was a very unlikely sum to have come from a real-world problem, and I naturally asked how it arose. I was told that it came from analyzing a queuing problem. If p(n) represents the probability of n customers in the queue, the recursion which the engineer derived for p(n) was

$$p(n+1) - 2p(n)/n - 3p(n-1)/n(n-1) = 0.$$

He had learned a method at the university for solving difference equations: Form a particular polynomial (in this case quadratic) using the coefficients in the recursion, find the two solutions of this quadratic, form sums of n'th powers of these solutions, and they are the general solution of the recursion equation. He needed to sum the series in order to find constants to satisfy the initial conditions.

He had memorized this method — and I was surprised that he had even heard of it — but he had no idea why and how it worked. Thus he did not know that it applied only to recursions with constant coefficients — while his equation had coefficients depending on "n". So he was proposing to apply a method which had nothing to do with the problem, and this would lead to nonsense. Then, if I hadn't stopped him, he would have summed the series on the computer — it converges very rapidly. As it happens, by multiplying the recursion equation by n!, you can convert it into an equation with constant coefficients, so that it can in fact be solved analytically. But that is not the main point of this anecdote.

3 When Bell Laboratories first acquired a large computer, it was under control of the Mathematics Research Department, which had the responsibility for providing this free service in the company. In order to help assure that the computer was being used wisely, we set up a committee in the department which screened all large and a sample of small problems. I was a member of this committee, and saw a lot of interesting uses. One example was an engineer summing a double series. It had only one fault: It was slowly divergent. Luckily we caught and stopped that one.

4 An engineer came to one of the mathematicians with a mathematical proof of an interesting new result. The mathematician rather quickly showed him a counter example to the proposed theorem. The engineer said "Oh, thank you! That's too bad. I am going to have to find a new proof."

5 Every year, the company did performance reviews of its employees in order to judge the quality of everyone's work. The usual experience was that it was very easy to agree on the top performers — the stars — and on the poorest performers, ones who had to show definite improvement during the next year or be in danger of being asked to leave. It was the large group in the middle, those not good enough to be stars but definitely an asset to the work, who were difficult to sort out. One year, a colleague of mine in the middle management ranks had a great idea. He established ten characteristic criteria of performance,

like quality of work, quantity of work, originality, cooperation, and keeping to the schedule, rated everyone from 1 to 10 on each performance characteristic, and added the ten number for each individual. He called me on the phone in great distress: This clearly identified the stars and the poor performers, but all the ones in the middle came out with scores between 45 and 55, and he knew that he could not trust the specific numbers to represent true differences in performance. What should he do? I had the pleasure of telling him about the central limit theorem, and the inevitability of his experience. One might even argue that some of the items, like quantity and quality of work, might tend to have negative correlation, so that the effect would be worse than for independent, identically distributed random variables.

6 In my early days with the company, I would get a visit every year or two from some engineer who was trying to approximate a discontinuous periodic function with a Fourier Series. He would find this strange little overshoot near the discontinuity, decide to take more terms, and make the overshoot narrower but just as big. What's wrong? Gibbs Phenomenon, which at that time was not as familiar to the engineers as it is today.

7 This brings me to the major internal education program which Bell Laboratories had established in the late 1940's for all incoming engineers without a master's degree. They spent three years, on the average half-time, learning linear algebra, complex variables, Fourier Series and Fourier and Laplace Transforms, probability theory, statistics, semi-conductor physics, and a number of other topics which at that time were not part of the regular university education of electrical and mechanical engineers. We taught the program ourselves internally within Bell Laboratories until the late 1950's, when some of the teaching was taken over by faculty from New York University. A few years later, these and related subjects had become part of the graduate and then the undergraduate program for engineers. The program was changed and we sent the incoming engineers to universities to obtain a master's degree in engineering or computer science. The internal education program became a continuing education system for essentially all employees - what you learned in school is never enough for a career in research and development. This exists up to the present time, with probably its major emphasis on computers.

8 One of the categories of employees in Bell Laboratories was that of technical associates, who were hired with degrees from two-year technical institutes and typically participated in laboratory or computer experimentation. A major institute from which we hired many of these people decided towards the end of the 1960's to expand from a two-year to a four-year program; for one thing, the government support for the education of veterans made this an attractive alternative. They took the existing two-year program, with its courses in electronics, circuits, etc., and added a second two years of partly fundamentals like calculus, basic physics, etc. Our opinion in Bell Laboratories was that this education was now in the wrong order, that the applications preceded the fundamentals, and we decided not to interview graduates of

this four-year program. After all, in a sense they competed with
regular four-year engineers, who had had their education in the RIGHT
order. The technical institute decided, not unnaturally, that if we
refused to interview the four-year graduates, they would forbid us from
coming to the school and interviewing two-year graduates as well! The
impasse was broken by the 1973 recession, during which we did no hiring
at all. By the time we were hiring again, the educational program had
improved. Incidentally, I am no longer sure that our position at that
time was indeed correct. Wouldn't the applied material make pretty good
motivation for the courses in fundamentals?

        9 Early in 1966, Bell Laboratories decided to broaden its
hiring pattern by looking not only for engineers (primarily electrical),
but also for mathematics majors who could become computer experts. We
had been hiring the engineers for many years and been giving them
additional education – as we discussed above; now we would also hire
mathematics majors and educate them further towards computer science,
and towards engineering. Our biggest problem was to be sure that very
bright mathematics majors, students who had grade point averages above
3.5/4 and had almost all A's in their mathematics courses, could
actually apply their mathematics to the real world. It is not possible
to be an A student in engineering and have no feeling for real world
problems, for applications to the real world are part and parcel of an
engineering education. But it IS possible to be a top student in
mathematics and still have no experience in, or talent for, applying the
mathematics to any other field. We looked for advanced courses in
science or economics or computer science as evidence; we looked at
summer jobs and hobbies; in short, at anything that might indicate an
ability to apply mathematics. This was very necessary.

        A View from One Employer

        What have I learned from experiences such as these over the
last 35 years? What do employers need from the mathematical education
of their staffs? At the elementary level, we particularly need the
following: The ability to set up the right problem, to have a good idea
how big the answer should be, and to get the right answer by ANY
available means whatsoever – mentally, calculator, paper-and-pencil,
computer, whatever. These abilities are needed of all employees from
secondary school graduates on up. At the more advanced technical
levels, we need employees who know that there is a large variety of
forms of mathematical thinking, and what these various forms can do.
Besides analytic, algebraic, and geometric thinking, there is, for
example, data-driven thinking, in which you must draw conclusions and
plan action on the basis of data, some of which are likely to be wrong.
There is probabilistic thinking, where the information is not
deterministic but stochastic. There is algorithmic thinking, where a
key point is evolution of an efficient procedure, probably on the
computer. There is the thinking which covers the areas of planning,
optimization, operations research, in which you consider defining an
objective, alternative ways of reaching it, the sense in which one
method is better than another, and what might be the best in the sense

of "best" which you have chosen. There is the combinatorial or graph
theoretic or number theoretic way of thinking - which we tend to call
"discrete" mathematics nowadays. Above all, we need the knowledge that
mathematical thinking, analytic, structural, quantitative, systematic
thought, can be applied to the real world and give valuable insights -
in other words, that mathematical modelling is possible and can be
successful.

It is also necessary to understand the mathematics! There
is a common caricature of an engineer as a person who looks up a formula
in a handbook, substitutes numbers, multiplies the answer by 10 (the
"safety" factor), and then builds it. Why does such a person have to
understand anything? Because memorizing the formula or the method and
turning the crank is fine for the standard text book, and for solving
yesterday's problems, but it simply doesn't work often enough in the
real world. Exercises at the bottom of the page in a mathematics
textbook usually practise only what is on top of the same page, but when
you meet a problem in the real world, you don't know what page it's on,
or even if it is in the book! Understanding is essential to applying
mathematics well. Walter Brattain, the Nobel-Prize-winning physicist,
explained in a talk which I heard that the key distinguishing feature of
the practise of science is the right to repeat an experiment, and
thereby to see for yourself that what has been claimed is actually true.
Of course the experiment may in some cases be too expensive, or too
dangerous, but the right to repeat it is there in principle. What
corresponds to this right in mathematics is the right to understand when
and how and why the mathematics works. It is essential for an employee
in industry.

Finally, a user of mathematics in industry must be prepared
for the open-ended. In education, mathematics is usually met in the
form of "here is a theorem, prove it" or "here is a problem, solve it".
In the real world, you must also be prepared for "here is a situation,
think about it. What do you think the problem, the right theorem, might
be?" Model-building courses are an especially good way for education
to prepare the future employee for this kind of activity. However, an
open-ended approach could be used in many places in the existing
structure of mathematics courses, and would be very vauable for the
student, as well as excellent pedagogy.

Mathematics in Telephony

Why did mathematics come so early to the communication
business in general, and to AT&T in particular? David Slepian, the
great Bell Laboratories mathematician, gave a talk a number of years
ago in which he philosophized about that. It is not only that
electricity is not easy to weigh or smell or touch (safely), so that it
leads naturally to more modelling and less direct perception. In
telephone communications, the behavior of the sender, the content of the
message, and the nature of the medium are all essentially probabilistic,
and hence require relatively sophisticated mathematical formulation.
Who will pick up the phone next, who will be called, will the call be

completed, when will the conversation start, will it be a person or a
computer or an image processor or a recording? All of these questions
can only be answered probabilistically. What will the message be? If
we knew, we wouldn't have to send it! (That is the key to the very
inexpensive mother's day messages in some countries: They look very
flowery and lengthy, but are in fact one of messages 1 - 16. The
information content is just 4 bits!) The length and content of the next
message can only be described probabilistically. The medium over which
the message is sent - cable, radio, satellite, optical fiber - distorts
the message in various ways, such as amplitude, phase, frequency
content. How can you describe the distortion? Again, only in terms of
probability. This is one reason that sophisticated applied mathematics
came to the telephone business as early as it did.

        A telephone system consists of transmission, of switching,
and of the operation of the entire system. Transmission has usually
been modelled using continuous mathematics, and modulation theory,
communication theory, and antenna theory, for example, were developed to
meet these needs. Nowadays, of course, transmission is often discrete,
and the mathematics classically associated with switching, that is
discrete mathematics, information theory, graph theory, and switching
theory also are seen to apply to transmission. The operation of a
communication system requires specific disciplines such as traffic and
queuing theory, and more generally all of operations research and much
mathematical economics. Truly many areas of modern applied mathematics
had their beginnings and development in Bell Laboratories.

Why We Teach Mathematics

        The total system for mathematics education has at least four
major purposes: The mathematics needed for everyday life, the
mathematics needed for intelligent citizenship, the mathematics needed
for your vocation or profession, and mathematics as a part of total
human culture. Traditionally, the mathematics of everyday life has been
the mathematics of the elementary school. The mathematics for
intelligent citizenship should basically be the mathematics of secondary
school - we shall return to this question shortly. The mathematics of
your vocation or profession is university mathematics if your profession
requires a university education; otherwise it is also a duty of the
secondary school. Mathematics as a part of total human culture has
really not been the responsibility of any level of education.

        Since the weakest part of this argument is without doubt the
mathematics for intelligent citizenship, let us return to it briefly.
What does this have to do with secondary school? The only secondary
mathematics that is ever claimed to have any aspect of intelligent
citizenship as its purpose is geometry, which is supposed to teach you
to think. Algebra and trigonometry, while certainly useful and also
valuable ways of looking at problems, are REALLY there to prepare for
calculus. What else is needed for intelligent citizenship? The ability
to reason from data, to handle probabilistic situations, to plan and
optimize and to understand something of modelling; to a lesser extent,

to think algorithmically, and discretely as well. These are many of the
same requirements as we saw for an employee in industry, but at a more
elementary level. I maintain, therefore, that probability, data
analysis, optimization, and modelling, are essential ingredients of the
school curriculum for all students – as is some familiarity with the
computer. This is true not because they are going to be scientists or
other employees of government or industry, but because they must
participate in the decisions which are the business of all citizens in
a democratic society. When should these subjects be taught? They can
perhaps begin in elementary school, but they are, in my opinion,
fundamentally the responsibility of secondary school – and for all of
our students.

        Why do we continue to hold meetings on mathematics
education? Because the four kinds of mathematics we have been
discussing – the mathematics for everyday life, for intelligent
citizenship, for your profession or vocation, and as a part of human
culture – continue to change. Why do they change? Because the
technology changes, because the applications of mathematics change, and
because mathematics itself keeps changing. All these changes find their
way back into the content and into the pedagogy of our total
educational system, and content and pedagogy must change in the light of
changes in technology, of applications, and of mathematics itself. We,
who concern ourselves with all of these aspects of mathematics and its
teaching, must understand the need for change and must provide some of
the necessary insights and leadership. Compare an international meeting
on mathematical research with one on mathematics education. You will
rarely hear a paper at ICM which could have been given 20 years earlier.
What has been said at this meeting in Udine which could not have been
said at Lyon in 1969, or Utrecht in 1967? There has been much that has
been new, and why was it new? Because it concerned the effects of the
changes in technology, in applications of mathematics, and in
mathematics itself that have taken place in recent years. These changes
are the fundamental forces that shape our view of why we teach
mathematics, including mathematics as a service subject.

Teaching first-year students

Fred Simons
Eindhoven University of Technology
Department of Mathematics and Computing Science
The Netherlands

There is a big difference between teaching undergraduate students having no idea why mathematics is included in their curriculum and what they can do with it, and teaching more specialised mathematical topics to graduate students who already know why they have included these topics in their curriculum and who have mastered the basic mathematical skills. Thus, for most teachers, the latter form a much more pleasant audience to teach. Service teaching to undergraduates has its own particular problems, both with respect to mathematical contents and to didactics.

In this contribution I will combine my own opinions with respect to content, presentation and developments in the near future with some of the points of view found in the contributed papers.

The starting point of the discussion document [4] for the ICMI-study is that mathematics is taught as a service subject in response to a need. It then concentrates on three questions:

- Why do we teach mathematics to students of discipline X?

- What mathematics should be taught to these students?

- How should this mathematics be taught?

I would like to start with considering these questions for service teaching to undergraduates, thereby describing the situation at this moment at many places. Next I propose to discuss whether the present situation gives a satisfactory answer to these questions in the light of the changing demands of the disciplines and the increasing availability of computer hard- and software.

Obviously, the answers to the questions "Why mathematics?" and "What mathematics?" depend on the need for mathematics in discipline X. Often this need arises from the conclusion that some mathematical tools are indispensable for solving problems in discipline X, and therefore these tools are wanted in the curriculum. If the use of mathematics in discipline X is not very widespread, this teaching usually suffices. For disciplines in which mathematics is used frequently, the demand of service teaching is usually considerably higher, both quantitatively and qualitatively; students in such disciplines not only need skills in applying mathematical techniques, but also an understanding of the mathematical concepts.

In any case, in the beginning of their study the students must be taught the elementary mathematical techniques. Therefore it is not surprising that all over the world undergraduate service courses mainly deal with calculus and linear algebra.

How these subjects should be taught depends on the need of discipline X. If the need is only an acquaintance with the tools, then teaching of the techniques will be sufficient. However, if the role of mathematics in discipline X is more fundamental, then already in the first year emphasis must be laid on the concepts as well. No doubt in the second situation the teaching should be done by mathematicians, but also in the first situation there are strong arguments for mathematicians being the teachers. They have the mathematical background for explaining mathematical techniques and a broader knowledge of which techniques can be applied in specific situations. Indeed, we see that with only a few exceptions the teaching of mathematics is done by mathematicians.

Since for many years the same calculus and linear algebra is taught nearly everywhere along more or less the same lines, it may seem that these courses provide a satisfactory answer to the questions raised in the discussion document. However, despite the striking similarity in these courses all over the world and despite our experience for many decades in teaching them, the situation is not as problemless as one might expect.

Roughly speaking, there are problems both with the motivation of the students and the attitude of departments and staff for the undergraduate service teaching. There are reasons to doubt whether our calculus and linear algebra courses are as useful as we think they are. The increasing use of new parts of mathematics and of computers in the various disciplines forces us to reconsider the contents of our undergraduate service courses. It might be that we have to conclude that the situation with respect to the undergraduate service teaching is not at all satisfactory and that we have to reflect on what really has to be done instead of what we do now.

Let us describe some of the problems in more detail.

In most mathematics departments research and the education of mathematicians have the first priority; service teaching is considered as a matter of secondary importance. Therefore often service teaching does not get the attention it needs. In particular when there is shortage of staff, as, for example, we were told is the case in Africa, the teaching of service courses is entrusted to young and inexperienced lecturers, sometimes graduate assistants, while the more experienced lecturers teach the honours or specialist courses.

Obviously this is a far from ideal situation. The importance of service teaching for a mathematics department has to be recognised. Often a mathematics department serves more students in other disciplines than students who major in mathematics and for that reason alone service teaching should not be treated as a matter of secondary importance. Good course management is essential if a department is to meet successfully the diverse objectives of the host departments, the sometimes less than enthusiastic staff and the large enrolments (Hodgson and Muller [3]). Service teaching on a major scale needs an organisation and management comparable with that of higher vocational schools, adapted to the university system. At some universities such organisations exist. For example, most of the engineering faculties in Japanese universities have special departments to teach mathematics to engineering students (Murakami [5]). Also at some universities in the Netherlands such organisations exist, but they strongly suffer from the governmental pressure on the departments to do research.

Many staff members dislike undergraduate service teaching. As Hodgson and Muller [3] formulate it, many mathematics faculty members perceive a service course as one or more of the following:

- large classes with a majority of uninterested students with a rather weak mathematics preparation,
- restricted and overloaded syllabuses, too difficult for the students and with topics remote from research interests, with emphasis on techniques,

-   a task for which little university credit is given.

Consequently, most teachers restrict themselves to the use of well-known standard texts and training in techniques; only a few are really interested in problems arising from undergraduate service teaching.

Now let us turn to how students may perceive a mathematics course. Since the students did not choose mathematics as their main field of study, they are tempted to give the subject a low priority and consider it just as a hurdle to jump over. Hence they are not motivated to spend much time on mathematics. Another reason for this attitude is that they have no idea about the role of mathematics in their discipline; lack of integration between mathematics and application raises doubts about the reason why some topics in mathematics are taught (Bacciotti and Boieri [1]). Many contributions to this conference contain similar complaints.

An interesting observation comes from Patetta [6]. He remarks that in countries where university study is considered to be a right for all the people (with adequate grounding), resulting in a non-restricted access to the universities, independent of the needs of the country, there often is an overpopulation in the classes with students without a definite vocation for their studies.

In order to improve the motivation of students, many contributions contain a plea for incorporating many examples from the student's discipline. However, it is not easy to find good examples. Most examples I have seen so far are either trivial, or need knowledge on the part of the student that is usually not acquired by the beginning of the first year or force the mathematics teacher to teach topics from discipline X. Since in general mathematicians do not have much knowledge of how mathematics is used in discipline X, close cooperation of the mathematicians and the teachers of the discipline will be necessary. In this way a better tuning of the mathematics courses and the courses of discipline X and the use of relevant examples in the math courses can be achieved.

The problem also exists on the other side. Often teachers from discipline X are scarcely acquainted with the contents of the mathematical courses and with progress in mathematical education. They sometimes use outdated mathematical methods and notation and are not aware of the mathematical equipment of their students. Related to this is the problem signaled by some contributors that many user departments have no clear idea about the mathematics they really need.

A final problem that I want to mention partially depends on the national situation. One of the problems a teacher may be confronted with when starting a first year course is the diverse mathematical knowledge of the group of students. For some disciplines the level may "automatically" be acceptable, e.g. physics, electrical engineering, but in other disciplines for some or for all students it might be necessary to start with secondary school mathematics, despite the fact that the students passed the school-leaving examinations. Some contributors to this conference mentioned the low level of preparation of secondary schools for university study.

So far I have concentrated on the existing situation and only mentioned some problems with respect to teaching and organisation. But matters are changing, and these changes cause problems too. The disciplines start to want to have more mathematics, but do not want to spend more time on mathematics in the curriculum. Combining this with a, to my opinion, decreasing level of knowledge and understanding on the part of first year's students, we are forced to teach more mathematics in a shorter time. But, as Siegel [8] formulates it, a student is not a vessel, we cannot just pour mathematics in. As was pointed out at Udine, a point is rapidly reached when the majority of the students cannot master so much mathematics in so short a time.

Particularly for disciplines that need a lot of mathematics, we have to admit that it is impossible to teach all the mathematics the students will or might need in their career. Indeed, often the problems in these disciplines are so complex that cooperation with applied mathematicians is necessary, and therefore ultimately the students must reach a level that they can communicate with these mathematicians (who, in turn, have to be able to communicate with people from other disciplines). In such situations the students need not only an ability to apply mathematics, but also a thorough understanding of the mathematical basic concepts. Teaching to these students must emphasise mathematical thinking and concepts rather than mathematical tricks.

For engineers, Roubine [7] suggests the teaching of mathematics more or less disconnected from other courses on the abstract level of that of today's papers, so that the engineer will be able to read what is published today or to discuss with mathematicians. Another important argument for this is that it is easier to acquire abstract knowledge at university, at the age of about 20, than later.

On the other hand, Bacciotti and Boieri [1] argue that the necessity for an engineering student to have a deep mathematical background is not always obvious. More precisely, everybody agrees that many sciences require very advanced mathematics on a research level. However, most graduates will find routine jobs, where they will be required only to use formulas discovered and verified by others.

Very interesting is the contribution of Clements [2]. In his opinion, the objectives for service courses to discipline X should not only be to teach the students a certain body of facts and techniques important in X, but also to teach the students how to acquire further knowledge and how to use their knowledge. He has made some successful experiments in this direction.

Now let us turn to the question whether out traditional first year courses indeed give the students any skill in applying mathematical tools or any mathematical insight. In Siegel's well-documented contribution [8] the situation in the United States is investigated with a very negative outcome. As she formulates it, the students seem to pass the examinations but they cannot do anything with the mathematics they have had. I have no reasons to believe that in other countries in general the situation will be very different.

One of the reasons for this bad result might be found in the interaction between the teaching system and the examinations. One of the goals of a service course is to provide the students with a certain number of techniques. These techniques can easily be assessed by means of short questions in standard examinations, and therefore this is the objective concentrated upon by most students and the majority of university staff. One of my chemistry students once explained her difficulties with studying mathematics. She said that the secondary school mathematics was simple; with every technique there was only one type of exercise, and conversely. Now she was confronted with simple problems where she had to choose which technique to apply. But she was never asked to explain why or under which conditions the technique could be applied.

Indeed, it seems that we have quite a lot of experience in teaching tricks to the students and asking back an application of the trick at the examination, but hardly any experience in teaching mathematical thinking.

Maybe this is a place to say a few words on "mathematical thinking" in a service course. Most mathematicians interpret this as "rigor", that no statement can be believed until a rigid proof has been given. However, very often proofs requiring a lot of ε–δ techniques are time-consuming and hardly contribute to understanding. I feel it very reasonable that a non-mathematician should believe a mathematician when the latter states that a certain statement holds.

In my opinion, a feeling for mathematics and mathematical sense is much more important than rigor. Even when teaching service courses at a technical and elementary level we can give the students a feeling for mathematics by not stressing the technique, but the heuristic considerations behind it. For example, one of my first-year engineering students argued that

$$\lim_{x \to \infty} \sqrt{x^2 + 4x} - x = 2$$

in the following way: for large x the graph of the root function is very flat, and therefore for large x the first term practically equals $\sqrt{x^2 + 4x + 4} = x + 2$, which implies that the limit equals 2. In my opinion he showed a very good understanding, but many of my colleagues disagree.

It is remarkable how many traditional exercises can be "solved" in a similar way by simply "looking at the formula". For example, the series

$$\sum_{n=1}^{\infty} x^n n^2 \log n$$

resembles a geometric series with common ratio x, since for large n the numbers $(n+1)^2 \log(n+1)$ and $n^2 \log n$ have the same order of magnitude. Hence the radius of convergence of this series must be 1.

The use of this type of heuristic consideration often already shows the correct answer to the problem before any technique is applied. In my experience it is very stimulating to the students to do things like this when lecturing. I am not sure whether these examples really demonstrate the teaching of mathematical insight, but this way of teaching certainly gives the students a feeling for mathematical sense.

The many complaints about the introductory courses concerning the inability of students to use the mathematics they have had and on a lack of conceptual mastery may raise severe doubts as to whether the traditional courses are as useful as we think they are. But not only for this internal reason is it absolutely necessary to reconsider the questions why we teach certain topics in what way; two external reasons are the changing demands for mathematics in the disciplines and the widespread availability of computers and software packages. The departments also beg for discrete mathematics, use of numerical techniques, difference equations etc. Indeed, at some places courses in discrete mathematics are already given in the first year service curriculum, and also at some institutions there are experiments with the use of the computer in service teaching.

However, I feel that we have to prepare for much more drastic changes in the traditional calculus and linear algebra courses. At the moment there is a lot of cheap and powerful numerical software available, and soon powerful computer algebra packages will be available too. Then the student will be able to solve nearly all the problems raised in the traditional courses and also much more complicated problems by simply pressing some keys of the computer.

Of course, it is not quite as simple as that. The students must be taught how to use the software packages and how to formulate the problem in such a way that they can solve them by using the packages. This requires a thorough understanding of the concepts, and I expect that it is on this aspect that the emphasis of future courses will be laid.

Many contributors stress the relevance of the use of computer algebra software. Siegel [8] formulates it in the following way: we would be remiss if we did not attempt to use symbolic manipulators in many service courses. They allow for exploration and experimentation with complex problems and, for non-mathematics majors especially, they represent the way they will be doing mathematics in the future. Those who have experimented report very satisfactory results. Many of the factoring problems, techniques of integration, complicated derivatives are quite irrelevant for these students. Frequently, they are more conversant with computers than is their mathematics instructor. Hodgson and Muller [3] remark that many know-hows conveyed in mathematical service courses have been readily available on computers for more than a decade, but that the impact on undergraduate mathematics teaching is still to be felt. They believe that it is vital that more mathematics departments begin experimenting with the introduction of symbolic mathematical software.

It is very encouraging to see that at those places which have started to experiment, the results are promising. However, these experiments are done by those few people who are really interested in the problems of service teaching or in the impact of the computer on teaching, and they all agree that a lot of research still has to be done. Unfortunately, most mathematics departments are not really interested in this kind of research; mathematical research is felt to be of much more importance than solving the didactical problems arising from the necessary changes in the service teaching. But maybe even a more serious problem is in the attitude of many of the experienced service teachers. Perhaps we are convinced that a lot has to be changed in service teaching, but most staff members are not, and stick to their traditional teaching despite the fact that the computer can do most of the things better, faster and more accurately than they can. They simply dislike computers, and argue that you do not teach the students mathematics when you teach them how to press some keys of the computer. Often they forget that for the students many of the techniques in the traditional courses are similar to computer programmes: they can be used without any idea why and how they work. Another argument often heard is that mathematics is mathematics, independent of the computer, and they are teaching mathematics. Indeed, I think that mathematics is the invariant in the traditional and future courses. But I expect that in the future the concepts will be central to the courses, while nowadays the techniques are the most important things. For example, now we concentrate on various techniques on finding a limit and hardly teach what a limit is; in the future we will concentrate on what a limit really is and how the concept of a limit or an approximation is used in various situations, while the computation of more complicated limits is left to the computer. The computer does not change the mathematics, but it changes the problems we are going to solve.

I believe that in future service courses will be both more fundamental and more practical than they are now, with a lot of non-standard, more thought-provoking problems and simple modelling, where for the more technical computations the students rely on the computer. To create such courses seems to me a challenge worth accepting.

References

[1]   Bacciotti, A. and P. Boieri: 'Teaching mathematics to engineers: the Italian case'. (See the Springer-Verlag volume.)

[2]   Clements, R.R.: 'Teaching mathematics to engineering students utilising innovative teaching methods'. (See this volume.)

[3]   Hodgson, B.R. and E.R. Muller: 'Mathematics service courses - a Canadian perspective'. (See the Springer-Verlag volume.)

[4]   Howson, A.G. et al.: 'Mathematics as a Service Subject', L'Enseignement Mathématique, **32** (1986), 159-172.

[5]   Murakami, H.: 'Mathematical education for engineering students in Japan'. (See this volume.)

[6]   Patetta, N.D.: 'Mathematics as a service subject - an interactive approach'. (See the Springer-Verlag volume.)

[7]   Roubine, E.: 'Reflections on the teaching of mathematics in engineering schools'. (See this volume.)

[8]   Siegel, M.J.: 'Teaching mathematics as a service subject'. (See this volume.)

TEACHING MATHEMATICS TO ENGINEERING STUDENTS UTILISING
INNOVATIVE TEACHING METHODS

R R Clements
Department of Engineering Mathematics, University of
Bristol, Bristol, BS8 1TR, Great Britain

Abstract. Teaching mathematics as a service subject makes
demands upon the teacher which are, in some ways, different
from teaching mathematics as a main subject and, in others,
similar but with an altered emphasis or level. This paper
first discusses the general aims of a degree course and then
identifies some ways in which traditional mathematics
teaching fails to meet these aims. Three innovations in
mathematics teaching, guided reading, simulation/case
studies and the continuous system simulation laboratory, are
then briefly described and the ways in which they contribute
to the achievement of the overall aims of a degree course
suggested. The strengths and drawbacks of each innovation
are mentioned.

INTRODUCTION
If we are to discuss the teaching of mathematics as a
service subject in tertiary education institutions it is appropriate
first of all to be clear what are the overall objectives of degree and
diploma courses in such institutions. It is the author's feeling that a
broad summary of the objectives for a course in X might be given as

  i)   to teach students a certain body of facts, techniques and
       principles important in X,
 ii)   to teach students how to acquire further knowledge and technique
       in X as and when the need arises subsequent to their formal
       studies and
iii)   to teach students how to use their knowledge of X in the solution
       of problems that arise in the real world.

It is not immediately clear to the author to what extent this broad
summary is applicable to the study of arts subjects but it is certainly
a reasonable statement when applied to the broadly vocational subject
areas - the natural sciences, engineering, mathematics, medicine,
economics, law etcetera. Accepting these principles it is then
reasonable to suppose that the mathematics instruction given to students
of these subjects should be guided by the same objectives. The author's
experience over the last fifteen years has been of teaching mathematics
to students taking degree courses in a variety of engineering
disciplines (including a degree course in engineering mathematics). The

three principles given above have guided his own development of ideas about the teaching of mathematics to engineers and have consistently been found to be applicable.

It seems unlikely that anyone would argue with the first of the objectives proposed. Indeed, since it is the objective most readily assessed by a conventional degree examination, the backwash effect of the examination system is to ensure that this is the objective concentrated upon by most students and by the majority of university staff. In many instances this concentration is so severe as to be detrimental to the pursuit of other possible objectives.

Whilst the relevance of the second objective is not new it has probably acquired increasing importance as the pace of technological and social change has quickened. Today it is even less likely than ever before that the knowledge and skills acquired at the beginning of a career will remain relevant and sufficient for a working lifetime. The ability to add new abilities to one's repertoire and adapt present skills in response to changing demands is very important. One cannot do this unless one has learnt how to continue to learn. This objective assumes additional importance when considering the teaching of mathematics as a service subject. Because of the pressures on the curriculum it is a common experience of mathematical educators that they are able to teach less mathematics than they consider desirable for the students concerned. Under these circumstances the teaching of the ability to learn mathematics independently is of vital importance.

For the teaching of mathematics as a service subject the third objective might be modified to read 'to teach students how to use their mathematical knowledge in the solution of problems that arise in the real world of X'. A common complaint of students learning mathematics in support of their main field of study is that the relevance of the mathematics is not clear. If they are taught not only mathematics but also how to use it in their own field of interest this complaint should be alleviated. This should not be taken to mean that only mathematics directly and immediately relevant to their main field of study should be taught. In many cases there are good arguments to be made for teaching some relatively abstract mathematics (for the sake, perhaps, of developing abstract reasoning and other skills). At least some of the mathematics taught must, however, be relevant to the main field of study and the use of such mathematics should also be taught.

The traditional methods of teaching mathematics both as a main discipline and as a service subject are much better suited to the achievement of the first objective than of the second or third. In the past teaching has been chiefly oriented towards communicating mathematical knowledge, technique and principle, and the learning of skills both of application and of further learning has been assumed to be largely incidental to the other activities during the degree course. Skills of application have been assumed to be acquired whilst learning about a series of standard mathematical models. The skills of learning, and general study skills, have been assumed to exist already, or to be

acquired naturally or, perhaps, to be fostered by personal tutorial
contact. In reality skills in these areas have been, at best,
incompletely learned by most undergraduates. That this is the case has
been well illustrated by reports from employers of graduates (Gaskell &
Klamkin 1974; Handelman 1975; Klamkin 1971; McLone 1973).

Over the period during which the author has been teaching mathematics
as an engineering service subject he has introduced or helped to develop
three particular innovations in teaching methods which have been
designed to improve, in areas addressed by the second and third
objectives, the abilities of graduates from certain courses at Bristol
University. To help our students learn how to learn mathematics
independently we have used the guided reading method to teach selected
courses. To develop the ability of our graduates to use mathematics in
the solution of real problems we have firstly used a technique best
described as simulation/case study and secondly developed a course in
continuous system modelling which is based around a laboratory format
using a specially designed continuous system simulation package and
hands-on use of microcomputers. The next three sections of this paper
briefly describe each of these developments in turn, identify their
strengths and disadvantages and suggest the remaining unanswered
questions and issues associated with these developments.

GUIDED READING

The educational effectiveness of the traditional technique
of university teaching, the course of one-hour lectures, has
increasingly come to be questioned. Bligh (1972), amongst others, has
extensively explored the merits and demerits of the lecture method.
University teachers have increasingly realised that a richer variety of
alternative learning experiences is both educationally more effective
and intellectually more rewarding for teacher and student.

Amongst the alternatives which have been proposed and explored is the
use of guided reading. The starting point for this departure is the
observation that, during a conventional lecture course, a considerable
part of the time is devoted to the transference from the notes of the
lecturer to the notepads of the students of relatively staightforward
factual material. When much of this material is readily available in
textbooks, it can be argued that the lecture time spent reproducing this
material is spent inefficiently or unproductively. Why not instruct the
students to read the textbooks instead? Most lecturers feel that they
have something in the way of insight, context and explanation to add to
the bare bones of the mathematical material presented in a textbook and
would therefore resist this course. Guided reading represents a halfway
house between the two extreme options. Given a textbook in which
sufficient of the factual and background material needed for the course
is to be found, the lecturer provides the students with a set of written
notes which specify, guide and supplement their reading of the book.
Their study of the textbook is then further supplemented by a series of
discussion classes or tutorials in which lecturer and students can
discuss, amplify and review the material being learned.

The theme of the 1977 University Mathematics Teaching Conference was teaching methods for undergraduate mathematics. In the chapter of UMTC (1978) dealing with guided reading a range of variations of the basic theme are described and reviewed. Amongst the merits of the guided reading approach mentioned therein are

- a) it develops students' confidence in their ability to read mathematical textbooks and learn mathematics independently of the lecturer,
- b) it reduces the sterile labour of note taking,
- c) it introduces more flexibility in the depth to which students study the material (thus enabling able students to be stretched without losing the attention of weaker students) and
- d) it generates greater student motivation and encourages students to discuss mathematical work amongst themselves.

The primary reason for its adoption by the author and his colleagues was the first though the others are of course welcome bonuses.

The prerequisites for a successful guided reading course are a suitable textbook and a set of instructional notes. (It should be mentioned incidentally that guided reading courses can also be framed entirely around material written by the lecturer but, in this case, the objective of teaching students how to use textbooks is lost and that was one of our primary aims). The notes require careful preparation. In courses prepared by the author such notes have functioned to

- a) specify the reading that was to be done,
- b) specify exercises to be done and provide solutions for student self-assessment,
- c) comment on sections of the book which the lecturer considers poorly explained or presented and provide alternative treatments or explanations,
- d) provide additional material omitted from the books and
- e) give guidance on notes to be made for reference and revision purposes.

The work set for the courses was divided into a set of units of roughly comparable work load. A series of discussion classes were also scheduled at which each unit's work was reviewed, discussed and set in context. It was made clear to the students that, in order to benefit from these classes, they must have completed their independent study of the material in the unit prior to the class. In the author's experience roughly 65% of the class contact hours which would have been devoted to lectures are needed for such discussion classes. Such classes should, ideally, take their direction from the lecturer and the students in roughly equal part. Experience indicates that it is a mistake for the lecturer to be purely reactive to the student demands in such classes - the lecturer must be prepared to challenge the students with new questions and ideas leading from their study of the textbook material. On the other hand sufficient time must be devoted to dealing with questions on both matters of understanding and principle and on the exercises set.

The student reaction to courses given by guided reading has been by and large favourable. Responses given by students informally can be summed up as :-
   a) Students enjoyed the courses. This was partly because of the novelty factor and the resulting increased range of educational experiences to which they were subject.
   b) Students would like more courses given in this style but not too many. The most popular estimate of the appropriate mix was approximately one third of all courses in the guided reading style.
   c) Students liked the additional freedom to pace their own work which guided reading gave, particularly the opportunity to work ahead when other work was slack.
   d) Paradoxically they also identified this freedom as a danger because they could easily defer work on these courses when other courses demanded more attention. They felt that some spur was needed to guard against undue procrastination. Work to be handed in from time to time might provide such a spur. The discussion classes were helpful in this respect.
   e) They reported that, initially, they were inefficient in taking notes from textbooks, copying large chunks of the book into their notes. Subsequently many students developed a style in which they read the material first, attempted the examples and then made much briefer notes. This style of note taking was judged to be particularly useful for examination revision.
To sum up then, the author and his colleagues have found guided reading to be an useful and enjoyable mode of teaching. It encourages students to develop their mathematical skills in ways which supplement the skills they acquire in more conventional teaching. The students appreciate the additional variety which it brings to their learning experiences. The technique is most useful when an appropriate textbook can be found for the course. Such a book must not only cover the appropriate material (or at least the major part of it) in an appropriate way but also be suitably priced for an essential student purchase. It is necessary for staff using this technique to prepare the course in advance and provide carefully written and structured supplementary notes. The class contact loading of the staff may be reduced by this technique but not very greatly. Approximate cost estimates made in UMTC (1978) indicate that guided reading is a more efficient use of staff resources but not by a very great margin. Care must be taken that the existence of a set of prepared supplementary notes does not prevent change for a number of years thus causing the course to become ossified. The method has been used at Bristol with groups of students up to about 20 in number. Reports in UMTC (1978) mention much larger groups though, in these cases, discussion in classes must be more difficult. One small problem which has been encountered occurred when one lecturer taught a guided reading course from his own textbook. Students reported that when, in classes, they asked questions about points in the textbook which caused them difficulties, they received the same explanation as was given in the book! This, under the circumstances, is not surprising. It does, however, indicate that lecturers should be wary of using their own books for a guided reading course and, if they choose to so do, take this

potential problem into account. A more detailed account of the
development and use of guided reading courses by the author and his
colleagues may be found in Clements & Wright (1983).

SIMULATION/CASE STUDIES

One of the principle deficiencies in the skills of
mathematics graduates entering employment in industry which is
identified in the literature, (Gaskell & Klamkin 1974; Handelman 1975;
Klamkin 1971; McLone 1973) is that they are lacking in the abilities to
recognise the mathematical properties of problems expressed in terms of
the (non-mathematical) problem source domain, to formulate a
mathematical model of the problem and, having solved that mathematical
problem, to re-interpret the mathematical solution into a statement of
solution in the problem source domain and to explain the solution, in
their own terms, to non-mathematicians expert in the source domain - in
short they have little ability in mathematical modelling. Whilst the
studies mentioned dealt primarily with graduates from mathematics degree
courses their findings should be equally applicable to the teaching of
mathematics as a service subject. Obviously service mathematics will
not teach the same level of mathematical skills but, within the
constraints of the level of mathematics taught, it is just as important
that students be taught not only mathematical skills but also the skills
of the creative application of mathematics to their own discipline. It
has already been pointed out that traditional university mathematics
teaching, and this includes service teaching, concentrates on teaching
mathematical skills and techniques, and that the system of assessment
generally used reinforces this tendency. In the case of service
teaching the restriction on the time available for mathematics also
causes lecturers to concentrate on getting over the mathematics to the
detriment of teaching its application.

The growing recognition of the shortcomings of traditionally
mathematically educated graduates has lead to a number of initiatives
both in mathematics degree courses and in service mathematics courses.
The main body of these have been in the area of mathematical modelling.
Initiatives in the teaching of mathematical modelling are described in
d'Inverno & McLone (1977), McDonald (1977), and Oke (1980) for instance.
There has been a parallel development of the philosophy of mathematical
modelling. The literature in this area was recently reviewed in
Clements (1982b). In the Engineering Mathematics degree course at
Bristol University a one term course has been introduced the aim of
which is to familiarise students with the problems and techniques of
using mathematics in the solution of real industrial and commercial
problems. The course is based round a series of exercises which are
best described as simulation/case studies. In these exercises students
work in small groups. This is, of course, a marked contrast with the
pattern of most of their degree studies where the emphasis is placed
upon individual work and responsibility. Group working and
responsibility is, however, typical of the working environment into
which most of them will go on graduation and some prior experience of
this is valuable.

At the start of each exercise each group of students is given a package of written material typically comprising reports, memoranda, design drawings, correspondence, data etcetera. These materials place the group in a simulated environment as mathematical practitioners facing a real problem. The problem is stated in the terms of the problem domain, not in mathematical terms, and it is the task of the students to grasp and understand the real problem, determine what sort of mathematics will help in its solution, develop a mathematical model of the problem and produce a mathematical solution, recast that mathematical solution in terms of the problem domain and finally evaluate what they have done and report it in appropriate terms. During this phase of the exercise the students are instructed to treat the members of staff tutoring the exercise as if they were the project leader or section head in charge of the work in the simulated environment. In turn the staff must play their role within this context. This will mean that the staff must react to students' suggestions and directions rather than imposing their own. They must be prepared to relinquish some of their control of the learning process. Their role becomes that of experienced practitioner offering advice and suggestions, exactly the role of the project leader or section head. Each exercise is designed to occupy two or three weeks. The students have one regular weekly meeting with the tutor in charge of the exercise and organise their own work in between meetings. They may, if they wish, consult the tutor more frequently but this is on an informal basis. This phase of the exercise constitutes the simulation aspect of the technique. It is a simulation in the same way as management games, military manoeuvres and other training exercises.

The production of the materials for the exercises could, possibly, have been undertaken within the University. However it was felt that, if it were possible to use real problems which real organisations had faced and solved, the realism of the exercises would be improved. Additionally, students would then, by the way, obtain some indication of the variety of challenging problems faced by industry and this might have the beneficial side effect of helping students in their career choices. A project was set up to approach industry and solicit suitable problems. The project, and the basic philosophy of the course, are described in more detail in Clements & Clements (1978). The simulation exercises actually used in the course are based on rather than directly copied from the industrial problems. In some cases the donor wished some changes to be made for commercial reasons and in most cases some changes were made for educational reasons. A side effect of obtaining problems in this way was that the method of solution adopted by the donor institution was available for comparison with the students work. This comparison is made after the students have finished their work on the problem and completed their report. It is in this way that the element of case study is introduced into the hybrid simulation/case study concept.

The course has been in use for eight years and sufficient experience has been accumulated to indicate the strengths and the drawbacks of the method. Firstly, how do students react to the demands made by this type

of work? Observation of students during the course and discussion of
the course with students at the end of the year have both contributed to
the identification of an overall pattern of student reaction. As might
be expected, students are initially bewildered and disoriented by the
open ended nature of the task and the lack of defined directions.
However, once they begin to succeed they rapidly gain in confidence and,
by the end of the course, most have developed at least some ability to
be creative with their mathematical skills. With the weaker students
the tutor often has to draw out ideas and help as unobtrusively as
possible in the formulation of the first models but once some measure of
success is achieved confidence usually grows rapidly. It is also
generally the case that, by the end of the course, students report that
the course is interesting and challenging and helps them to see the
relevance of the mathematics they are learning.

The problems posed by assessing student performance are, as may be
surmised, different from those posed by more conventional courses. A
technique which averages a series of subjective assessments made by the
staff tutoring the exercises has been evolved. The method is described
in more detail in Clements (1982a). The author does not, in fact,
consider assessment of the course particularly vital. One of the
purposes of assessment is to motivate the students. Motivation has not
generally been found to be a problem on this course – the fascination of
the problems is usually sufficient to guarantee student involvement.
The principle benefit derived from the course is the changed attitudes
and approaches of the students. Such changes, being more in the
affective domain than the cognitive one are difficult to assess by
formal means.

The primary drawbacks of the simulation/case study method of teaching
are the demands made upon staff by the unconventional role which it
calls upon them to adopt and the staff contact time needed. The method
obviously calls upon staff to play a role, a somewhat different way of
relating to students than their usual one. Within the role staff must
react to student suggestions in an appropriate way. There is virtually
no scope for didacticism in this situation. Tutors must be prepared to
embark upon each session with each student group with no certain
knowledge of where the session will lead. More than anything this
demands that tutors have the confidence to put themselves in an exposed
position. The method is also somewhat demanding of staff contact time.
As used at Bristol each group of three students has a weekly
consultation with their tutor which is nominally three-quarters of an
hour. We have not used the method with groups larger than 20 students.

This description of the simulation/case study method has been
necessarily briefer than might be desirable. More detailed description
of the technique and some examples of the materials used may be found in
Clements & Clements (1978) and Clements (1978, 1982a, 1984a, 1988). A
description of a similar course developed independently at Oklahoma
State University, USA, may be found in Agnew & Keener (1980, 1981) and
Agnew et al (1983). The author is also aware of similar developments at
Lulea University, Sweden.

THE CONTINUOUS SYSTEM SIMULATION LABORATORY
The simulation/case study course described in the last
section has provided some training for our undergraduates in the general
area of mathematical modelling as well as much experience in other
skills which feature in the post-graduation working environment.
However, as far as the mathematical modelling is concerned, the course
rather throws the students in at the deep end. Other initiatives
(d'Inverno & McLone 1977; McDonald 1977; Oke 1980) have approached the
teaching of modelling in a somewhat more structured and gradual way.
One aspect of modelling which the author and his colleagues have
attempted to teach in a more structured way is the computer simulation
of continuous and discrete systems. Note that, in the last section, the
term simulation was used to denote an instructional method. In this
section the term is used in the description of the material and
techniques being taught. The two uses should not be confused.

The role of simulation in mathematical modelling courses is well
established (Huntley 1984; Moscardini et al 1984). Modelling exercises
and activities often result in models which lack viable analytical
solution techniques. In these circumstances simulation offers a
solution route (and one that would be adopted in an industrial or
commercial environment). Simulation may be implemented either by
writing a specific computer program designed for the problem under study
or by the use of one of a range of general purpose simulation systems
such as GPSS, CSMP, CSSL, ACSL, DYNAMO, TUTSIM and many others. The
major languages like CSMP, ACSL, CSSL, GPSS and DYNAMO are usually
available on multi-user mainframe or mini computers. The author's first
approach to teaching simulation used one or other of these languages.
It was rapidly apparent that students at first found such systems far
from simple to use although regular users quickly gained adeptness.
Further, most multi-user computers in institutions of higher education
are heavily loaded with resulting poor response, at least during class
hours.   As a result of both these factors students who needed to
simulate a fairly simple system preferred to write their own simulation
program for the problem in hand using a language with which they were
already familiar (usually Pascal, Fortran or Basic). As microcomputers
became more widely available in the university such programs were
increasingly written using these machines, again usually for reasons of
convenience and familiarity and particularly because an increasing
number of students owned their own microcomputers. The conclusion must
be that, whilst simulation should be a regular tool of the mathematical
modeller whether student or experienced practitioner, in practice the
main available large systems are not ideally suited to the needs of
tertiary education courses in modelling. This conclusion is reinforced
by the experience of Moscardini et al (1984) who describe a simulation
package, IPSODE, written within their institution specifically as an
introductory simulation package for their students.

More recently simulation languages for microcomputers have begun to
appear. Experience with the TUTSIM package, a block oriented

continuous system simulation language available on Apple microcomputers, indicated that students found it particularly easy to learn and use. The block oriented input language was appealing, particularly to engineering students who are familiar with control theory ideas, in its visual and diagrammatic approach to the construction and representation of models. The package also has particularly good facilities for the graphical presentation of results. The course at Bristol had to be based in a laboratory equipped with Acorn BBC microcomputers. Since there was no suitable simulation system available for this microcomputer the BCSSP system (Clements 1984b, 1985a), which uses largely the same input language and provides similar facilities to TUTSIM, was written. BCSSP has powerful facilities for the presentation of results in graphical form including the comparison of results from different simulations and can rapidly produce hard copy of graphical output. This facility was felt to be important for its role in the systems course. The availability of this package has opened up possibilities for new teaching styles which are being exploited in the teaching of systems modelling.

One of the skills which is felt by the author to be important to the effective user of simulation is the ability to integrate analytical and theoretical study of a system with numerical investigation of that system via simulation. Without that skill there is a severe danger that the user of the simulation package merely accumulates mountainous quantities of simulation data about the system under various conditions without any theoretical framework within which to organise and understand that data. The importance of this ability is not, of course, limited to simulation. The capability of integrating analysis and numerical work is important for the effective user of mathematics generally and as such contributes to the satisfaction of the third major objective suggested above. The availability of the BCSSP package has made possible the design of a course aimed at helping students develop this skill through a kind of mathematical laboratory.

The main objective of the laboratory is to provide a series of investigations which demand that the students integrate simulations and the application of their more theoretical analytic skills in order to elucidate the properties of some systems with which they are relatively unfamiliar. This synthesis of investigative modes is firstly a very powerful exploratory technique and secondly one which, in the author's experience, undergraduate students need more assistance in developing. The essence of the synthesis lies in the realisation that professional mathematicians do not make theoretical advances in a vacuum. It is more often the case that mathematicians know (or at least have a hunch about) what they want to prove before the proof is sought. That knowledge may come from any one of a variety of sources but numerical studies and simulation are a common and powerful source of such pre-knowledge. The abuse of simulation is to use it as a device for avoiding having to grapple with the theory behind a problem. Whilst studies of systems may sometimes unavoidably be based solely on simulation it is, in general, not the optimal mode of investigation.

The type of problems which have been sought for this module are problems in which, initially at least, analysis does not provide a complete understanding. Progress may perhaps be made by exploratory simulation which reveals the principle features of the system. Theoretical explanations for the revealed behaviour may then be sought, often using techniques such as linearisation, small order or large order analysis in some parameters etcetera. If such explanations are found their appropriateness and range of validity may then be checked by further simulation. Obviously not all problems will exhibit all these features, but the course overall is intended to develop the abilities of the students to use both simulation and theoretical analysis in ways which support and illuminate each other. The lecture content of the course introduces theoretical ideas, such as linearisation and stability, which underpin the use of simulation in this manner. The philosophy and implementation of the course is described in more detail in Clements (1985b, 1986).

It is only possible, of course, to use a method such as this if appropriate computer resources are available. The laboratory used by the author's department for this course allows students access to one microcomputer per student pair. If larger groups of students were to take this course it would be necessary to split them into smaller groups and run parallel courses or to expand the laboratory. Either option has serious resource implications. The course also relies heavily on students undertaking computational project work between formal laboratory sessions. This work is not, in our case, timetabled but it does, of course, require that the students have adequate access to the laboratory facilities between formal sessions. That, in turn, requires that the laboratory have spare capacity and is not so heavily used for formal teaching that informal student access is very restricted. The requirements for laboratory backup to this teaching are therefore fairly stringent.

The problems which are used in the laboratory also need to be carefully selected and prepared. A modest library of such material has been developed by the author and his colleagues but there is a need for more. Mechanisms for the cooperative development and exchange of such material between institutions is very valuable.

The examination of the course is less easy than conventional courses. Ideally it would be desirable to be able to set a form of practical examination with students using microcomputers and the simulation package to carry out an investigation of an unfamiliar system using the techniques which they have learned during the course. Currently resource and other technical difficulties have prevented the implementation of such an examination but the concept has not been discounted for the future. For the time being a conventional written, three-hour, unseen examination is used.

CONCLUSION
   This paper has described three innovations in teaching
methods which have been adopted in the teaching of mathematics to
engineering and engineering mathematics undergraduates. In each case
the innovation has been driven by the possibility of meeting a wider
educational objective than is readily feasible using the traditional
lecture method. Each innovation brings with it strengths and
weaknesses, problems which must be overcome for effective implementation
and use. It is the experience of the author and his colleagues that the
effort to do so is worthwhile when evaluated in terms of the additional
skills which can be engendered in the students by the broader range of
learning experiences resulting. Teaching mathematics as a service
subject is a challenging activity. The rewards of an innovative
approach both in personal satisfaction for the teacher and improved
learning for the student are considerable.

REFERENCES
Agnew J L & Keener M S (1980). A case study course in applied
          mathematics using regional industries. Amer Math
          Monthly,87,55.
Agnew J L & Keener M S (1981). Case studies in mathematics. UMAP
          J,2,3,1981
Agnew J L, Keener M S & Finney R L (1983). Challenging applications:
          problems in the "raw". Maths Teacher,76,274.
Bligh D (1972). What's the use of lectures?. Harmondsworth: Penguin.
Clements R R (1978). The role of simulations/case studies in teaching
          the practical application of mathematics. Bull
          IMA,14,295.
Clements L S & Clements R R (1978). The objectives and creation of a
          course of simulations/case studies for the teaching of
          Engineering Mathematics. Int J Math Educ Sci
          Technol,9,97.
Clements R R (1982a). Initial experience in the use of
          simulations/case studies in the teaching of Engineering
          Mathematics. Int J Math Educ Sci Technol,13,111.
Clements R R (1982b). The development of methodologies of mathematical
          modelling. Teaching Maths Applics,1,125.
Clements R R & Wright J G (1983). The use of guided reading in an
          Engineering Mathematics degree course. Int J Math Educ Sci
          Technol,14,95.
Clements R R (1984a). The use of the simulation and case study
          technique in the teaching of mathematical modelling. In
          Teaching and Applying Mathematical Modelling, ed. Berry J S
          et al, Ellis Horwood.
Clements R R (1984b). BCSSP (Bristol Continuous System Simulation
          Package) - User Manual. Department of Engineering
          Mathematics, University of Bristol.
Clements R R (1985a). BSETR (Bristol State Equation Translator) - User
          Manual. Department of Engineering Mathematics, University
          of Bristol.

Clements R R (1985b). The development of a mathematical modelling skill
        through the use of dynamic simulation. In Proceedings of
        the 2nd International Conference on the Teaching of
        Mathematical Modelling, Exeter Univ, July 1985.
Clements R R (1986). The role of system simulation programs in teaching
        applicable mathematics. Int J Math Educ Sci
        Technol,17,553.
Clements R R (1988). Case studies in industrial and commercial
        applications of mathematics. Cambridge University Press (in
        preparation, to be published in 1988).
Gaskell R E & Klamkin M (1974). The industrial mathematician views his
        profession: a report of the committee on corporate members.
        Amer Math Monthly,81,699.
Handelman G H (1975). The mathematical training of the non-academic
        mathematician. SIAM Review,17,541.
Huntley I D (1984). Simulation - its role in a modelling course. In
        Teaching and Applying Mathematical Modelling, ed. Berry J S
        et al, Ellis Horwood.
d'Inverno R A & McLone R R (1977). A modelling approach to traditional
        applied mathematics. Math Gaz,61,92.
Klamkin M (1971). On the ideal role of an industrial mathematician and
        its educational implications. Amer Math Monthly,78,53.
McDonald J J (1977). Introducing mathematical modelling to
        undergraduates. Int J Math Educ Sci Technol,8,453.
McLone R R (1973). The training of mathematicians. SSRC Research
        Report.
Moscardini A O, Cross M & Prior D E (1984). On the use of simulation
        software in higher educational courses. In Teaching and
        Applying Mathematical Modelling, ed. Berry J S et al, Ellis
        Horwood.
Oke K H (1980). Teaching and assessment of mathematical modelling in an
        MSc course in mathematical education. Int J Math Educ Sci
        Technol,11,361.
UMTC (1978). Possible models for guided reading courses. In Teaching
        methods for undergraduate mathematics, Proceedings of the
        University Mathematics Teaching Conference 1977. Shell
        Centre for Mathematical Education, University of Nottingham.

DISCRETE MATHEMATICS: SOME PERSONAL THOUGHTS

J.H. van Lint, Department of Mathematics and Computing
Science, Technological University Eindhoven,
The Netherlands.

Contrary to what one would think, judging from the recent
stream of textbooks on "Discrete Mathematics for Computer Scientists",
this subject is not the union of all subjects in mathematics that are
not in the calculus course, but necessary for computer science. This
note will try to show what discrete mathematics really is.

In the description below we will assume that students have had some
calculus, linear algebra (maybe even elementary probability theory)
and know what mathematical reasoning is. A course in logic could
precede or follow the discrete mathematics; it is not part of it.

We will concentrate on principles, ideas, and the way of thinking which
are the essence of discrete mathematics. The following diagram
illustrates the comments below.

FINITE STRUCTURES

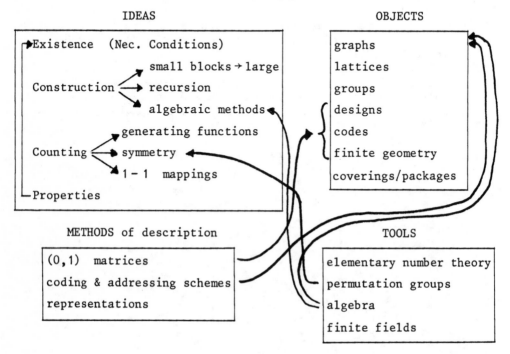

The main topics in the study of finite structures are <u>existence</u> questions, <u>constructions</u>, <u>counting</u>, and also studying <u>properties</u> of the objects in question.

One should stress the occurrence in <u>many</u> different situations of similar principles, e.g. in construction one has:

    a)   using several <u>small</u> objects to construct one large one.
    b)   recursive constructions.
    c)   using algebraic techniques to construct combinatorial objects.

In counting one uses:

    a)   generating functions (either ignoring convergence, questions or cleverly using them),
    b)   symmetry principles (e.g. permutation groups),
    c) $1-1$ mappings of seemingly different objects onto each other.

There are many combinatorial or just "finite" objects to study. Some are mentioned in the diagram. The "tools" are themselves objects of study in discrete mathematics.

In conveying ideas and ways of thinking the "methods of description" are what is important. They are sometimes no more than <u>bookkeeping</u> devices but one should try to teach how to use and <u>choose</u> them in such a way that they are an aid in solving the problems one is interested in. E.g. a <u>clever</u> way of <u>numbering</u> or <u>addressing</u> the vertices of a graph can solve many questions on <u>paths</u> in this graph. The <u>representa-tion</u> of a combinatorial object in <u>other</u> terms is often half of the solution of the problem one is interested in.

Tools such as <u>permutation groups</u> play a rôle in questions of counting, e.g. what does it mean to say that two objects are essentially the same? Many parts of algebra (groups, rings, ideals, boolean algebra) play a major rôle. One of the most important is the theory of <u>finite</u> <u>fields</u>. At present, many combinatorial objects can be constructed <u>only</u> (with sometimes a few exceptions) using finite fields.

In DM courses one can use (from the start or eventually):

<u>algorithms</u>, principles of <u>optimization</u> and one should give many <u>applications</u> to a variety of subjects.

Where does one find applications, respectively fields of education, requiring parts of the above? There are many, e.g.

statistics (quality control): designs
electrical eng. (communication): codes, boolean algebra
computer science: graphs, algorithms, (0,1)-matrices
business administration: graphs, algorithms
social sciences: graphs.

Examples

Many of the subjects taught can be illustrated by examples which most students find fascinating. These examples provide strong motivation. Personally, I prefer giving them _after_ the relevant mathematics has been treated. I mention a few:

(i)   Graph-addressing

In some systems a message goes through a telephone network preceded by the _address_ of the destination. At each vertex the message is directed along an edge that brings it closer to the destination. The problem of giving each vertex an address from $<0,1,*>^n$,   such that distance of addresses equals distance in the graph (* does not contribute) leads to very interesting problems.

(ii)   Switching problems in communication.

(iii)   Hadamard matrices were used to transmit the pictures of Mars made by Mariner '69. It takes less than an hour to show how the quality of the pictures was increased tremendously by the use of an _error-correcting_ code.

(iv)   There are many interesting examples from _quality control_ etc. of the use of Latin Squares and Block Designs.

(v)   _Write once memories._ (Exercise for the reader!)

Several memory systems for computers (punched cards, compact disks) cannot be reused (a bit = 1 is a hole that is there to stay). Can one design systems to reuse them nevertheless? Here is an example that the students like very much, given after treating something that maybe looks useless.

One wishes to store one of the numbers 1 to 7 on four successive occasions. This can be done with a 12-bit memory. On each occasion three bits are used (to store 1 to 7 in binary). However, it can also be done using only a _seven-bit_ memory. One uses the so-called Fano plane (7 points, 7 lines) as in the figure below:

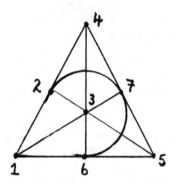

(The set {2,6,7} is also a line!)

On the first usage a number, say 5, is stored by punching a hole in
position 5. How to proceed on the next three times this memory is
used? The reader should try it.

     (vi) Conference telephone calls
           Electrical engineering students will easily understand the
requirements of an electrical network (without resistances) that makes
it possible to have a telephone conference with n persons. Each
person should be able to hear each of the others equally good (no
energy loss, etc. etc.). It takes only a few minutes to translate the
requirements into the following:

Is there an n by n matrix C for which all diagonal elements are
0 , all the others have the same absolute value (say they are ±1)
and such that any two rows of C have inner product 0 ? In other
words:
$$CC^T = (n-1)\ I\ .$$

These matrices (called conference matrices (!)) occur in a chapter on
designs.

     (vii) Search time for data stored in a computer.
           We store data having n properties each of which can be
one of two kinds (yes = 1, no = 0) . The data is stored in batches
(or bins, = bin packing). E.g. if all data with first coordinate 0
is in one bin, this bin gets the name (0***...*) . One can show that
in order to minimize worst-case search time the list of names of the
bins is a matrix of 0's, 1's and *'s such that each row (resp.
each column) has the same number of *'s , every column has some fixed
number of 0's (resp. 1's) and each sequence in $\{0,1\}^n$ belongs to
exactly one bin. Clearly a problem from design theory.

     (viii) Winning in a football pool
           In a football pool one can forecast the outcome of a number
of matches (often 13), namely:

     0 = draw,   1 = home team wins,   2 = visiting team wins.

If all 13 forecasts are correct one gets first prize. To guarantee
this one needs 3^{13} forecasts (expensive!).

Suppose one aims for second prize (i.e. one error). This can be
achieved easily by handing in 3^{12} forecasts. However, it can in
effect be done by only 3^{10} forecasts!! For this, one uses coding
theory, namely the perfect ternary Hamming code of length 13 .

(This idea is still not economically practical but there are practical
schemes designed by using these ideas.)

(ix) Pictures of Mars (Mariner '69)

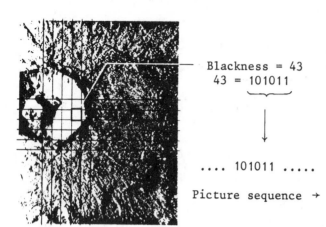

Blackness = 43
43 = 101011

.... 101011 .....

Picture sequence →

A picture is divided into little squares (pixels). For each pixel the degree of blackness is measured (in a scale of 0 - 63, in binary). So, this degree is described by a sequence of six 0's and 1's. The picture results in millions of 0's and 1's to be transmitted to earth.

The transmitted message is corrupted by noise. The effect is that some 0's are interpreted as 1's (and vice versa). As a result the quality of the picture could become extremely bad.

Suppose we are willing to take roughly five times as long to transmit the message. We could repeat each bit five times. This would lead to a substantial improvement but nowhere near to what was achieved in practice. The following solution was used:

A Hadamard matrix of order n is an n by n matrix H with entries ± 1 such that $HH^T = nI$. (These play an important rôle in Combinatorics, turn up quite often and are part of many courses). Construction is easy if n is a power of 2 (induction).

Consider the 64 rows of the two matrices H and $-H$ (H of order 32). To send the number i, transmit row number i. The receiver takes the received message $\underline{x}$ and calculates $\underline{x} H^T$. If there are no errors, then the result has 31 coordinates 0 and one coordinate equal to ± 32. Now, suppose there are t errors ($t \leqslant 7$). The coordinate that should be ± 32 still has absolute value $\geqslant 18$, all others should be 0 but aren't. However, they have absolute value $\leqslant 14$. So the correct value of the darkness can still be established.

In practice, it was extremely unlikely that a sequence of 32 signals contained more than 7 errors. Result: beautiful pictures.

MATHEMATICAL EDUCATION FOR ENGINEERING STUDENTS

Haruo MURAKAMI
Department of Applied Mathematics
Faculty of Engineering
Kobe University
Kobe 658, Japan

Brief Review of Mathematical Education in Japan

There are at present 95 national, 36 provincial and 334 private universities in Japan. Most of them have Engineering Faculties for research into engineering and education of engineering students. University education is for four years following on from six years primary, three years middle school (junior high school) and another three years high school (senior high school) education. There are also many junior colleges with two years courses for subjects like English literature, home economics, nursing and so on. For engineering students, however, two years education is considered to be too short, and there are 58 technical colleges offering five years education for middle school graduates.

Normally, after completing compulsory education at the end of middle school, one goes to an ordinary high school to study general subjects like Japanese, English, social sciences, mathematics, natural sciences such as physics, chemistry and biology for three years. There are engineering, commercial and agricultural high schools as well. Those who would like to learn specialized skills go to these vocational high schools, but the demand is small and the number of these high schools is not large. Altogether, over 95% of students go on to high school. High school education often seems to be a part of compulsory education. On graduating from high school, most students like to go on to higher education, and the entrance examination to universities becomes very competitive.

The mathematics subjects taught at ordinary high schools are as follows: numbers and expressions, equations and inequalities, functions such as quadratic, fractional and irrational functions, and plane analytical geometry in the 1st year. Algebra and geometry for second year students contains conic sections, two dimensional vectors, 2 by 2 matrices, and 3 dimensional vectors and geometry. Second year students also learn fundamental analysis, which contains series, functions such as exponential, logarithmic and trigonometric functions, and introduction to elementary calculus. Calculus and probability and statistics are given to third year students. The former contains limits, differentiation and its application, and integration and its application. The latter contains permutation, combination, binomial theorem, probability, probability distribution, descriptive statistics, and statistical induction.

Many high school students in Japan have to work very hard. Since
their main concern is to pass the entrance examination for universities,
they study how to attack and solve given problems, and usually do not
enjoy mathematics.

Mathematical Education in the First Half of a University Course

The main mathematical subjects taught in the first half of
university education are calculus and linear algebra. Teaching is done
by staff belonging to the Faculty of Liberal Arts, i.e. Junior College.
What other subjects should be taught is left to the lecturers of the
Faculty. Discrete mathematics such as set theory, abstract algebra and
mathematical structures are being introduced in some universities. One
thing which should be mentioned is that students do not usually work hard
in this period since they are almost sick of studying after all the hard
work preparing for the entrance examinations. Instead, they want to
enjoy university life. Another reason for their not studying hard in
this period is that they do not yet realize how important and useful
mathematics is for their later study.

Mathematical Education in the Latter Half of a University Course

Usually students go into senior college after one year and half or
two years of junior education. The length of time they spend in junior
or senior colleges depends on the university.
Standard subjects taught to senior engineering students are
functions of one complex variable, special functions such as Bessel
functions, ordinary and elementary partial differential equations,
numerical analysis and perhaps classical vector analysis. Discrete
mathematics containing mathematical logic, set theory and graph theory
are beginning to be introduced. Subjects like functional analysis,
advanced numerical analysis, probability and statistics, abstract algebra
and advanced partial differential equations are taught to first and
second year postgraduate students.
One special feature of the Japanese educational system for
engineering students is that most of the Engineering Faculties in
Japanese Universities have special departments, called Kyoutsuu Kouza
(servicing departments), to teach mathematics to engineering students.
The number of staff belonging to these departments differs from
university to university. Roughly speaking, there are more staff in
universities with more engineering students and fewer in those with fewer
students. Usually, there are staff in the pure mathematics department of
the Faculty of Science who are teaching their own pure mathematics
students aiming to be mathematicians or school teachers, although a large
number of them take jobs such as system engineers. Since staff belonging
to the Faculty of Science and the Faculty of Liberal Arts (junior
college) do not usually teach mathematics to senior engineering students,
responsibility for teaching mathematics to engineering students for the
latter half of their courses falls on the staff in the mathematics
servicing department in the Engineering Faculty. If the department does
not have enough staff to teach its own students, it usually employs
several part time lecturers from other neighboring universities.

Survey of Staff Opinions

To investigate how to improve the mathematics education for engineering students, a questionnaire was recently sent to all the staff of the engineering faculty of the university where the author works and also to staff of mathematics servicing departments in major universities. Each questionnaire contained a list of mathematical subjects and a series of boxes corresponding to each subject. The respondent was asked to mark the boxes to indicate whether or not he or she thought the subject should be taught, how it should be taught (i.e. putting stress on the acquisition of knowledge and skills of mathematics or on the mathematical thinking), by whom (i.e. by mathematicians or by engineers), and using which methods. The methods suggested are the top down way of teaching in which general theory is taught first and examples are explained by applying the general theory, and the bottom up way of teaching in which examples are given first and general theory is introduced by extracting properties common to these the examples. Respondents were given space to write free opinions.

The results obtained from this questionnaire showed first of all how many staff considered that each of the selected subjects should be taught. Essential, important, desirable and unnecessary subjects were defined as those where more than 75%, 50 to 74%, 25 to 49% and less than 25% staff in the indicated field of engineering or servicing department considered that the subject should be taught. Calculus, transformations, and linear algebra subjects (with the exception of standard forms) were regarded as essential or important by all the engineers and mathematicians. Information theory and mathematical linguistics subjects were largely seen as unnecessary, although a few of the subjects were seen as desirable by chemical engineers and information and system engineers. Discrete mathematics was generally seen as desirable, although there was a large range of opinions according to the individual subject and specialism of the respondent. Complex functions and differential equations were evaluated very differently by different specialists. They were generally seen as essential or important, with the exception of chemical engineers and electrical and electronics engineers (although the latter thought complex numbers and complex functions to be essential). A similar pattern was observed for vector analysis and numerical analysis subjects, although some subjects were considered to be important by all specialists. Architectural and civil engineers considered every field of numerical analysis to be essential. On the whole, functional analysis and control theory were seen as desirable or unnecessary, although mathematicians, electrical and electronics engineers, and information and system engineers showed the most enthusiasm. Chemical engineers thought that all the subjects in these fields were unnecessary, and information and system engineers considered Hilbert spaces to be essential. Control theory was also seen as either unnecessary or only desirable. The one exception to this was that numerical control theory was rated as important by chemical engineers. Probability and statistics were seen as relatively important. As expected, information and system engineers considered all the subjects in this field essential, and with the exception of analysis of variance, so did architectural and civil engineers. Mathematicians generally saw these subjects as important, as

did the other engineers, although analysis of variation and statistical
estimation and tests were only desirable.

For the second question, how mathematics should be taught, the
results of the survey are as follows. Both engineers and mathematicians
in the servicing departments think stress should be on the acquisition of
skills and knowledge of mathematics in subjects like calculus, complex
variables, Fourier analysis, differential equations, vector analysis,
numerical analysis, and probability and statistics. The only subjects
where mathematical thinking should be stressed more than knowledge are
discrete mathematics and functional analysis. For linear algebra and
differential equations, different opinions were expressed for different
part of each subject. One interesting result here is that mathematicians
in the servicing departments are more convinced than engineers that
skills and knowledge are more important than mathematical thinking for
engineering students. (Different opinions would probably be heard if pure
mathematicians in the pure mathematics departments were surveyed.)

As for the question of whether mathematicians or engineers should
teach mathematics to engineering students, the result of the survey is a
little different from what was expected. Since most of our engineering
staff tend to consider themselves as those not having enough mathematical
knowledge, it is understandable that they responded that mathematicians
should teach most of the subjects. The result we did not expect is that
a fairly large number of mathematicians in the servicing departments
responded that some areas of applied mathematics would be better taught
by engineers. Both engineers and mathematicians responded that calculus,
linear algebra, discrete mathematics, complex functions, Fourier
analysis, differential equations, vector analysis and functional analysis
should be taught by mathematicians, and numerical analysis should be
taught by engineers. The only difference is about probability and
statistics and mathematical control theory. Engineers thought these
subjects could be better taught by engineers, whereas mathematicians
thought the other way round.

The results obtained from the question on the method of teaching
show that both engineering staff and mathematicians in the servicing
departments consider that both the top down and bottom up methods
suggested should be employed, depending upon the subject being taught. It
seems to be natural that they think the top down method is better for
teaching fundamental subjects like calculus or theoretical subjects like
functional analysis, and the bottom up method is better for teaching
applied or practical mathematics such as numerical analysis.

Improving Mathematical Education for Engineering Students

A basic question may be raised here. Is mathematics for engineering
students essentially different from that for pure math major students ?
Should the way to teach mathematics to engineering students be different
from that to pure math students ? Mathematics for engineering students
should not be very different from that to math major students. One might
say that mathematics for engineering students is after all mathematics.
For the second question also, one may answer that there should not be
much difference in teaching ways. But one should be aware of the fact
that pure math students are those who are already attracted or enchanted

by the beauty of mathematics, and they are willing to study mathematics
as their major subject, whereas engineering students think of mathematics
as merely a tool to use for their engineering studies.

As mentioned above, engineering students do not realize the
importance and usefulness of mathematics while they are studying
calculus, linear algebra and so on. It is hard for them to appreciate the
necessity of studying it. Later, when they need mathematics, they find it
difficult to catch up, either because they are too busy with studying
proper engineering subjects or because of their lack of fundamental
knowledge of mathematics.

How can we solve this problem ? One good solution for this may be
to give the students motivation to study. How can they be given
motivation ? Teach with examples. For instance, consider the way that
ordinary differential equations are taught to 3rd year engineering
students at Kobe University. First of all, they are given an example
from population dynamics. The equation given first is

$$\frac{dy}{dx} = a\,x \quad .$$

As is well known, this simple Malthus model is not very good at all.
They are next introduced to the famous Verhulst model:

$$\frac{dy}{dx} = a\,x\,(k - x)$$

By giving this model, the lecturer can talk about asymptotic
behaviour of a solution, equilibrium points, and stability of a solution.
Solution curves can either be drawn by hand or shown by using a computer
display in the classroom.

Then, a small parameter c is added.

$$\frac{dy}{dx} = a\,x\,(k - x) - c$$

This parameter c may correspond to capturing animals in Africa, or to
deforestation by human beings. The parameter c is then increased a
little. Here the lecturer is in a position to be able to talk about the
structural stability. The parameter can then be increased a little more.
If the parameter c is increased still further, then, all of sudden, the
structural stability is lost, and students realize that all the solution
curves drop away from the top left to the bottom right. The lecturer can
then explain what environmental capacity means to us, and how easily we
can destroy our world.

This example is very useful if the lecture is for general students.
If all the students are engineering students, it may be better to pick
examples from engineering which are equally as good. If the student is
in the electrical engineering department, it is best to take examples
from electrical engineering. Teaching mathematics with examples like

these will convince the students that mathematics is useful and
important. To do this, the instructor should know the engineering as well
as the mathematics. Does this mean that mathematics should be taught by
engineers ? Although it is not necessarily true that mathematicians can
teach mathematics better than engineers, it is certain that the more
advances are made in the engineering, especially in areas dealing with
computers, the more sophisticated mathematics and mathematical thinking
will be needed, and mathematicians can definitely better at teaching them
and conveying a feeling for mathematics to students. So, perhaps the
subjects should be taught by mathematicians with the help of engineers in
deciding what examples to pick, in what way to teach, how to construct
the teaching method, and so on.

When mathematics is taught by mathematicians they tend to teach it
with too much stress on the mathematical rigorousness or teach subjects
which are interesting only from the theoretical point of view. Sometimes
they try to make students fully understand things such as epsilon delta
arguments or how to complete an incomplete infinite dimensional metric
space. These arguments would be necessary for pure mathematics students
but not for engineering students. On the other hand, engineering
students should be told of the existence of a nowhere differentiable
continuous function, although giving an example with the full proof may
not be necessary. What should be taught to engineering students is, in
addition to the mathematical knowledge, mathematical thinking, a feeling
for mathematics and mathematical sense or "a sense of mathematics" like
"a sense of humour". If mathematics is taught entirely by engineers, it
may be hard to convey those essentials to students. Thus fundamental
mathematics and abstract mathematical subjects are probably best taught
by mathematicians in a well prepared, well organized and appropriate way
for engineering students with the help of engineers.

Considering the method of teaching, the best way is perhaps to give
a few typical examples to motivate the students and then extract
properties common to these examples to introduce a general theory, and
then, after proving theorems, more examples of applications should be
given to show how powerful the theory is. This combines both bottom up
and top down methods.

Influence of Computers

The effective use of new technology must be considered. Once upon
a time, man used only his own strength, but he then learned to use the
strength of domestic animals such as horses to do his labour. Later,
engines were developed which were hundreds or thousands of times more
powerful than the animals. These engines now routinely do work that
would be nearly impossible by manpower alone. Life without these
machines is virtually unthinkable.

In the same way, man has also learned to use technology for his
intellectual work. Computers have been developed which can process
information hundreds or thousands of times faster than the human brain,
and routinely do work that would be nearly impossible by brainpower
alone. There are still many areas where computers cannot replace the
power of human thought, but they are now becoming essential tools, and
life without them is nearly unthinkable.   Think of the fact that the

contents of several volumes of an encyclopedia can be stored with dynamic pictures in one compact disk. Mathematics education must be changed to use the benefits gained from the progress of computers. Database and artificial intelligence techniques such as knowledge engineering will be of great help for education. Numerical analysis in its present form is already impossible without using a computer.

At the very least, we can install a microcomputer in the classroom with a big screen to demonstrate graphically the locus of functions, solution curves of differential equations and so on [1]. This can be very effective. Afterwards, we can ask students to do their assignments either by using their own personal computers or by using computing facilities provided by the university.

A computer algebra system can be employed as a very powerful tool in the mathematics education. Quite recently, the symbolic manipulation system Reduce was implemented to run on MS-DOS so it can be used on personal computers. Until now, muMATH has been used on micros, but it was thought that bigger machines were needed for better computer algebra systems such as MACSYMA or Reduce. Now Reduce is available on micros, and soon there will be reasonably good symbolic manipulation systems on pocket computers. In a few years time, engineering students will carry computer algebra systems in their pockets, like they carried slide rules thirty years ago. Life has changed, and so the subjects which should be taught must also be changed. How to integrate rational functions by using partial fractions is still taught, but teaching subjects of this sort may no longer be necessary. This has already been discussed in the ICMI study [2] and elsewhere [3], so it will not be repeated here.

References

1   Murakami, H. (1985). A View of Computer Assisted Mathematical
            Instruction with Particular Reference to Simulation and Use
            of Graphics in Analysis Courses, Invited Lecture, Proc. 3rd
            Southeast Asia Conference on Mathematical Education

2   Murakami, H. & Hata, M. (1986) Mathematical Education in the Computer
            Age, The Influence of Computers and Informatics on Mathema-
            tics and its Teaching, edited by A.G Howson and J.-P. Kahane,
            ICMI Study Series, Cambridge University Press, pp.85-94

3   Murakami, M. Hata & T. Yamaguchi (1987), A New Way of Teaching
            Mathematics Using Computer Algebra Systems, Microcomputers in
            Secondary Education, Elsevier Science Publishers B. V. North
            Holland pp.395-400

SOME REFLECTIONS ABOUT THE TEACHING OF MATHEMATICS IN ENGINEERING SCHOOLS

E. ROUBINE
(Retired) Professor at University of Paris and Ecole
Supérieure d'Electricité - France

This paper presents a few reflections resulting from a long experience teaching at the University and Engineering schools (Ecole Polytechnique, Ecole Supérieure d'Electricité) and industrial consulting. These reflections will probably appear unconventional and, may be, a little bit provocative. We shall therefore first indicate the limits of this presentation : it is strictly limited to the French system of education and to the only technical field familiar to the author, that is electronics in its modern day meaning, including the usual electronics of components and circuits, communications and radio and a large part of computer science and control theory. These correspond, for instance, to the scope of the French "Société des ingénieurs électriciens et électroniciens"or the American I.E.E.E.

Electronics is a field where education presents an especially acute problem in that the field has an amazing multiplicity, an accelerating evolution and resorts to an extensive mathematical background as signal theory or electromagnetics.

Choosing which branches of mathematics to teach becomes a real headache if one intends to cater for short term as well as for medium and long term needs.

The short term :
The short term is essentially concerned, within the University and Engineering schools, with mathematics that can be of use to the teaching of other fields and to the beginning engineer. At least in the French system the reality is often disappointing between the following extremes.

One extreme is the case of a curriculum corresponding closely to the often conservative demands of the physics or technical faculty : the everlasting special functions, operational calculus,..., many recipes, a "tool-box" mathematics rather than mathematics as a tool.

This education greatly suffers from the absence of important pedagogical simplifications brought by theories largely taught elsewhere (Lebesgue integral, Hilbert spaces, Schwartz distributions, exterior calculus,...).

At the other extreme, communication between teachers becomes so difficult or idle that it results in a disconnection where the mathematician ends up in a separate world and a situation usually considered as absurd.

The medium and the long terms :
The real problem is as everyone knows, what mathematics to
teach to the aspiring engineer (or physicist) which he will need in
his professional life in a 5 to 10 years perspective (medium term).
Beyond (long term) any forward planning is often bound to be defeated.

In his day to day activity, even when it is technical, the
electronical engineer apparently needs only very little mathematics.
The slide rule of yesterday or today's pocket calculator seem to be
largely sufficient. In fact, especially in research areas, the compo-
nent/system dialectics brings in a distinction between "components-
engineers", more inclined towards physics, and "systems-engineers",
who lean more towards calculus and whose mind is more closely oriented towards mathematical structuralism.

In the area of development, one cannot ignore the modifications introduced by C.A.D. For example, the design of V.L.S.I.
components is faced with hard algorithmic problems which is the
realm of the systems-engineer.

The relation of the engineer with mathematics is elsewhere.
An important part of the engineer's activity is permanent self-improvement
through the reading of technical publications and participation in
conferences. This requires in many cases a fairly strong mathematical
background. The authors are young and use a modern language. It is
reasonable to consider that the engineer needs before all mathematics
as a communication language, whereas it is generally admitted that he
should be taught mathematics as a tool. By the way one can emphasize
the fact that a good mathematical culture as very useful even if the
engineer does nothing but to use, in the computer, the results of published papers. Calculations can be shortened and frequent mistakes
avoided.

Applied mathematics.
A common perception is that mathematics for the engineer is
the so-called applied mathematics. Old books covered fields largely
forgotten now, such as descriptive geometry, nomography, graphical
statics. Today numerical analysis, probability and statistics are
frequently taught as applied mathematics even if the first is
grounded on functional analysis and the second on measure and integration.

In fast evolutive techniques, like electronics, it is not
easy to determine the contents and the limits of a specific curricumum.
One merely applies "pure" mathematics to the problems at hand. Who
could have foreseen the use of Galois fields in coding or, in other
domains, the use of number theory in cryptography, of algebraic
topology in the chemistry of large molecules? Even more striking is the
example of linear integral equations of first kind. They were of
little interest to mathematicians. The conditions under which they
can be solved are restrictive, but, above all, they come into play in

ill-posed problems as defined by Hadamard, i.e. problems whose solution is extremely sensitive to fluctuations of the data (such as
numerical approximation or experimental noise). This corresponds in
particular to the so-called inverse problems -to go back from effects
to causes- which are extremely important in numerous areas : geophy-
sics, radiology, optics, electromagnetics, metrology,... The pressing
needs of the users have thus led mathematicians to come back to
these equations and technically acceptable methods of solution have
been proposed.

As a last example one could mention, among the fashionable
theories, those referred to as of maximum entropy, used to solve
problems of spectral estimation and extrapolation, commonly used in
signal theory ad image processing (antennas, seismology, radiology,
tomography, N.M.R., radioastronomy,...).

It is interesting to note, in this respect, one of the
salient aspects of modern electronics. The use of more and more
elaborate (at least to the engineer) mathematics is made possible by
technical advances, microelectronics providing the circuits that can
process increasingly faster larger and larger amount of data. Data
processing has thus largely invaded experimental physics and most
contemporary techniques. An example is the modern viewpoint on noise
reduction. Nowadays people deal with clever coding schemes, i.e. by
mathematical craft in connection with microelectronics. Space communication, videodiscs, compact discs are among numerous applications.

All in all, one can observe that how to apply mathematics
to ever new problems is, more and more, a matter for true professionals,
something generally not addressed in Engineering schools. One could
say that, at least in the field of electronics, there is no longer
applied mathematics but mathematicians whom the engineer engaged in
research or development will eventually consult.

One is led to reshuffle some accepted views : formal
mathematics used as a communication tool by the ordinary engineer, but
transformed into an actual tool by a professional mathematician, possibly the engineer himself if he has the taste and the talent. It
should be noted that many advances in electromagnetics, signal theory,
information theory, have been achieved by mathematicians (compare
the Nobel prizes in economics).

The teaching of mathematics :
The above development tends to show the usefulness of a
sufficiently high-level mathematical culture, so that the engineer
can read professional literature and discuss with the mathematician he
consults. Hence the idea that mathematics, even as a service subject,
taught at the University or in Schools should be essentially used,
in the medium term, as a communication language.

This teaching will be part of the basic training, and it is
not absurd to see it as more or less "disconnected" from other
branches which take advantage of the initial training of the students.
In the French educational system, most of the students of the
engineering schools originate from excellent preparatory classes

("Mathématiques spéciales").

These considerations would lead to a somewhat reduced basic course, modern and of high level, consisting essentially of functional analysis (where numerical analysis itself has its foundations, e.g. the fashionable finite elements method) and probability theory.

The other professors can spend some lectures on review of their own mathematics, with whatever language and notations are best suited. In fact, this is often the situation. Calculus of variations is presented in control theory, a part of probability and statistics, including random functions and detection theory are conveniently taught with signal theory. However the case of finite mathematics is to be considered apart. Even if connected to a computer science or a communication course it is today one of the major topics to be taught.

The notion of a fairly autonomous course with the level of abstraction corresponding to an advanced content and to a modern presentation is not easily adopted by the college administration, the faculty and, to a lesser extent, the students.

A word regarding abstraction in the so-called modern mathematics and the permanent attacks from an older generation of engineers and experimental physicists. One should bear in mind that this is only a matter of epistemology. "Modern" mathematics is only apparently more abstract than older mathematics. The approach of mathematics as a service is not to practise some "concrete" mathematics but to substitute an abstract model for the concrete physical world, a model for which the language of mathematics makes sense. This is not a modern idea. The aim of applied mathematics is to find appropriate models for the real world.

Lastly, we would like to present efficiency arguments of a psycho-pedagogical nature. The Montessori method (after the famous Italian educator) proposed, some 75 years ago, to bend children's education according to their sensitive periods, i.e. their areas of interest and facility. In a similar way one can consider that twenty-year old brains are more amenable to abstract notions and that this should be exploited. At 30 or 40 to learn new mathematics "in abstracto" is arduous as opposed to experimental physics or even more technical fields which require more maturity in the understanding of reality. One could think of such mathematics as an investment made with the optimal return. However it is to be recalled that in continuing education where people have a solid professional experience they are highly motivated to handle mathematics as a service. Nevertheless, do not forget the law of the French psychologist Th. Ribot (1839-1916) concerning the gradual degradation of the memorization duration with the age.

Computers.
The importance of computers is too widely known to be brought up again, except to stress the following evolution in the

engineering world. Computers are so powerful that one is led to resort
to more and more complex theories as soon as they are applicable. The
example of integral equations is a case in point. They used to appear
only in very advanced curricula. Fredholm's theory, originating in
problems of potential theory, was not easily solvable through numerical
methods. It is of interest to mechanical or electrical engineers and
has already been used by Poincaré for antennas. With the computer,
integral equations are now familiar to eventual users. One could mention some research in electromagnetic diffraction with singular
integral equations, finite parts and distributions processed by
computer.

An epilogue by way of conclusion.
Are discussions about pure or applied mathematics, mathematics as a culture factor or a service, ... so necessary? Every one
does his best in his sphere. This brings back to me a short story
heard from the French physicist Abragam. In a circus there was a
funambulist on a bicycle in equilibrium on a tightrope. He had a
violin on which he tried to play the Kreutzer Sonata. Then someone in
the public whispered in his neighbour's ear: "Well, last month I heard
Yehudi Menuhin in the same piece, and he was very much better".

TEACHING MATHEMATICS AS A SERVICE SUBJECT

M.J. Siegel
Mathematics Department, Towson State University,
Towson, Maryland 21204, USA

Mathematics has served science and commerce for thousands of years. Yet the attraction of mathematics to mathematicians has frequently been the pure beauty of the subject without regard for its applications. Hence, for many teachers who first found mathematics magnetic for its own sake, the acts of teaching and having to justify the service function of their courses is often a challenge. Even now, many American graduate students in mathematics are exposed only to the relationship of mathematics to physics. Their graduate training does not require even a nodding acquaintance with statistics, operations (operational) research, model building, numerical analysis or simulation techniques. Since those students who are trained in the most abstract mathematics are the most likely to be teaching in the nation's colleges and universities, American undergraduates face a ridiculous irrelevancy in the mathematics classroom.

In most American colleges and universities today incoming students can choose a first mathematics course from

- remedial mathematics - intermediate algebra
- finite mathematics - usually for business students
- liberal arts mathematics - "math for poets"
- precalculus - functions (algebraic and transcendental)
- technical mathematics - for the trades (two year colleges)
- statistics - for social science, business, economics and health sciences
- short calculus - for non math and science majors
- calculus for mathematics, science and engineering- three or four semester sequence in which advanced placement is possible
- discrete mathematics - for computer science and mathematics students

In the US in 1980 of all students registered in elementary (but not remedial) courses at the level of calculus or below, only about 30% were enrolled in the calculus for mathematics, science and engineering (Conference Board of the Mathematical Sciences 1981). The overwhelming majority of students in beginning mathematics courses are enrolled in other service courses required by their choice of major. Proper placement in courses (by type and level) is particularly important since the mathematics course is used as a filter (implicitly or explicitly) in many disciplines.

An awareness of the problems being faced by mathematical educators in the 1980's led to the convening of two important conferences. Both of these were funded by the Alfred P. Sloan Foundation and became a springboard for many of the projects which will be mentioned here. The first conference, held at Williams College (Massachusetts) in the summer of 1982, featured talks on the first two years of college mathematics. The proceedings, published by Springer-Verlag (Ralston & Young 1983), make for excellent reading. Mathematics as a service course is treated in at least half of the papers. The second conference, held in the summer of 1984, was the first national meeting on mathematics in the two year colleges. Two year colleges in the US have several missions. Among them are the education of technical workers, who generally go directly into the work force, and the preparation of the more academically inclined who will enter four year colleges and universities to complete their education. The Springer-Verlag publication of those proceedings (Albers et al 1985) is similarly useful to those who are interested in the improvement of mathematics education.

The Mathematical Association of America (MAA) has a standing committee on service courses (of which I am a member). Its purpose is to explore for the American mathematical community the mathematical needs of other disciplines. It is revealing to note that this committee is a subcommittee of two of the most active committees of the MAA - the Committee on the Undergraduate Program in Mathematics (CUPM) and the Committee on the Teaching of Undergraduate Mathematics (CTUM). The first of the parent committees recommends curriculum, while the second attempts to aid in the effective delivery of the material to the student. These two aspects of mathematical education enter into all the deliberations of the Committee on Service Courses. Its most recent project was a three year study on the need for discrete mathematics in the first two years of the (American) undergraduate mathematics curriculum. I chaired the panel which recently published a 104 page report, available for a small charge from the MAA. Supported by the Alfred P. Sloan Foundation, the panel was made up of representatives of the computer science community, engineering schools and applied and pure mathematicians.

The study grew out of the dissatisfaction of many outspoken computer science educators who felt that the offerings of mathematics departments to computer science students were inappropriate in content and in timing. Of course, there was a forceful reaction by mathematicians who strongly resented the attack on the calculus - the mainstay of the freshman mathematics diet. What was revealed by our panel's research was that engineering faculty, chemists, physicists and biologists were dissatisfied with the calculus as it was being taught and begged for more discrete methods and techniques. They said that students lacked conceptual mastery of the subject and could not apply what they had learned. They asked that the mathematics of the first two years be more

integrated - the continuous and the discrete presented side by side to
show the relationship and effectiveness of the two. Many asked that
difference equations and recurrence relations with a heavy emphasis on
techniques of problem solving and the use of the computer be included in
the standard curriculum.

It became quite clear that what had begun as a study into the needs of
computer science students had quickly evolved into a critical review of
the entire elementary mathematics curriculum traditionally prescribed
for all science students. The panel was not prepared to recommend
revolution but stongly urged that discrete mathematics become a part of
the first two years' offerings and that, to allow room for it, there
should be careful attention paid to the restructuring of the traditional
three semester calculus sequence. Widespread dissatisfaction with the
mathematical maturity and skill of students who had gone through the
traditional courses indicated to the committee that something was amiss.
We had many debates as to whether it was the material, or the way it was
taught; whether it was the student or the system. The calculus books
had gotten easier since the 1960's. The students were getting less and
less out of a more elementary syllabus. Our mission was to make
recommendations about discrete mathematics and yet we felt that we
needed to know more about how people learn and how abstract ideas,
generalizations, and algorithmic analysis should be taught. We
submitted the report with some misgivings, knowing that with so much
more to learn about our failures with teaching calculus (more than 200
years in the maturation), it was presumptuous to pretend we knew how to
introduce beginners to discrete mathematics. The Committee on Discrete
Mathematics in the First Two Years (Report 1986) recommended that:

1.   Discrete mathematics should be part of the first two years
          of the standard mathematics curriculum at all colleges and
          universities.

2.   Discrete mathematics should be taught at the intellectual
          level of calculus.

3.   Discrete mathematics courses should be one year courses which
          may be taken independently of the calculus.

4.   The primary themes of discrete mathematics courses should be
          the notions of proof, recursion, induction, modeling and
          algorithmic thinking.

5.   The topics to be covered are less important than the
          acquisition of mathematical maturity and of skills in using
          abstraction and generalization.

6.   Discrete mathematics should be distinguished from finite
          mathematics, which as it is now most often taught might be
          characterized as baby linear algebra and some other topics
          for students not in the "hard" sciences.

7.   Discrete mathematics should be taught by mathematicians.

8.   All students in the sciences and engineering should be
            required to take some discrete mathematics as undergraduates.
            Mathematics majors should be required to take at least one
            course in discrete mathematics.

9.   Serious attention should be paid to the teaching of the
            calculus. Integration of discrete methods with the calculus
            and the use of symbolic manipulators should be considered.

10.   Secondary schools should introduce many of the ideas of
            discrete mathematics into the curriculum to help students
            improve their problem-solving skills and to prepare them for
            college mathematics.

In January, 1986 a conference on calculus was convened in New Orleans,
LA. The conference grew out of the observation by many mathematicians
that calculus courses had grown to be ineffectual and irrelevant. The
discrete mathematics study had shown that this was a widespread belief.
Conference participants were divided into three workshops: Curriculum,
Instruction, and Implementation. The report from the conference, Toward
a Lean and Lively Calculus, is available from the MAA as Number 6 in the
MAA Note Series. The recommendations of the conference are sweeping,
covering major changes in the number of topics in the syllabus (a
reduction with more time allowed for non-standard, thought-provoking
problems), more applications, heavier reliance on computer software and
computer-based instruction, suggestions for small-group and discussion
sessions with smaller classes and new textbooks. Discrete methods and
numerical techniques are considered central. Since one-third of the
conference was focussed on instruction, the final report contains a
large section on what we know about teaching and learning mathematics
and how to improve both.

In recognition of the importance of the issues brought to the fore in
the work of the discrete mathematics and the calculus groups, a CUPM
Subcommittee on the First Two Years of College Mathematics has been
formed. That committee will concentrate on courses beginning at the
discrete mathematics - calculus level. Among the tasks of the committee
is the identification of the subject matter and skills that are
important for the students at this level. What are the proper sequences
of courses for students in various majors? What is the place of
geometry (planar, analytic, spatial) in the curriculum? What is the
role of computers in instruction at this level? Does calculus provide
the best foundation for students in the disciplines we serve? How can
we attract students with the highest potential to continue to study
mathematics? How can we best write a two year syllabus to combine
continuous and discrete mathematics for students in the mathematical
sciences? How can we apply what we know about cognition to the
construction of syllabi?

I would like to share with you some of the information now available to the Committee on Service Courses of the MAA. We have been trying over the last few years to be in direct contact with the professional organizations of the disciplines which we perceive to be the heaviest users of mathematics courses. From some of the disciplines we get enormously enthusiastic response and from others practically nothing. Each member of the committee has adopted a discipline. Since mine was computer science, I got involved with the discrete mathematics project. Our most useful replies from other fields have been from biologists, chemists and the engineers.

Our survey of sixty engineering departments has found that, in general, most of the departments were pleased with the quality of the mathematics service courses offered to their students and reported that almost all mathematics departments were cooperative and willing to discuss mutual concerns. They expressed some dissatisfaction about the amount of mathematics taken by engineering students. That engineering students do not elect to take any more than the bare minimum in mathematics may be a reflection of some of the respondents' observation that mathematicians do not seem to have a favorable attitude toward the engineering students. There was, again, the plea that theory must be taught so that the student will be able to apply the concepts. Changes in the notation from mathematics to engineering courses cause students a great deal of confusion. That is because there seems to be a lack of retention of the basic cognitive scope of the mathematics courses. It was suggested that we might look at each other's textbooks to see how our students have to cope with different approaches. Also revealed in the survey was a dissatisfaction with the knowledge and ability of students to handle geometric ideas (analytic, plane and spatial). There was a suggestion that we put more emphasis on visualization. The low level of preparation in most American high schools was mentioned by almost everyone. Mathematics departments at the university level are accused of not correcting the deficiencies. With respect to what should be taught besides the standard calculus and differential equations, many mentioned complex numbers, numerical methods, discrete mathematics, geometry and linear algebra. To get a copy of the report, contact Professor David Ballew, Western Illinois University, Maconb, IL 61455.

What do chemists need in mathematics? Harvey Bent, Chairperson of the American Chemical Society's Committee on Professional Training recommends that chemistry students know more discrete mathematics, more linear algebra, more statistics, but - and he stresses this - they should be able to use the mathematics that they have already been exposed to. They may have been taught the mechanics of calculus, but cannot handle applied problems even after several semesters of calculus. He recommends to his students that they take an engineering mathematics course so that they will have the experience of actually doing lots of applied problems. The American Chemical Society(ACS) has a Division of Computers in Chemistry. In a course they recommend for undergraduates they include topics such as capabilities of digital computers, accuracy

and precision, significance of numbers, error accumulation,
floating-point number systems and design of algorithms. A large portion
of the syllabus is devoted to numerical methods: integration, solution
of differential equations, matrix manipulations, solutions of linear
equations, combinatorics, pseudo-random number generation for
Monte-Carlo calculations, modeling and simulation, quantum chemistry,
reaction dynamics and statistical mechanics. There are other topics in
data collection and experimental design which involve both statistical
and mathematical notions (Fourier transforms and convolutions, etc.).

Computer Applications in Chemistry is a junior- senior level course that
has been taught at The Pennsylvania State University for ten years
already. All students are required to know FORTRAN. The topics are
similar to the ACS course just described. In particular, there is a
major emphasis on numerical algorithms (similar to the content of a
mathematical numerical analysis course) including curve fitting,
function generation, numerical integration and solution of differential
equations. A section on eigenanalysis: finding the characteristic
polynomial, solving for its roots and solving sets of linear equations
for eigenvectors. Monte Carlo methods and Fourier transform methods are
also in the syllabus. An interesting nonnumerical topic that is
included is graph theory. The purpose is to show that the connection
tables have a mathematically rigorous underpinning and that graph theory
theorems can be profitably exploited in chemistry. [A connection table
involves representing a structure by a matrix or set of matrices where
the entries represent bonds or atom types.] Examples of the use of the
graph isomorphism theorems abound in the handling of chemical structure
information handling and substructure searching. Optimization methods
are discussed including the simplex method applied to gas chromatography
and flame spectroscopy. There is some attention paid to pattern
recognition, artifical intelligence and information retrieval.

Professor Peter Lykos of the Illinois Institute of Technology and an
active member of the ACS Committee on Professional Training spent
several hours with the MAA Service Course committee. He pointed out
that the traditional structure of the chemistry curriculum has four
emphases; physical, organic, inorganic and analytical. Although we
associate mathematical methods primarily with physical and analytical
chemistry, there are important ways in which mathematics is used in
organic and inorganic chemistry.
He stressed that:
  1. Students need to learn to think mathematically and
      learn to handle notions of symmetry, probability etc.

  2. There has been a resurgence of classical mechanics and
      solid geometry with important applications.

Beyond the standard calculus, he recommended training include
instruction in

      1. error assessment and propogation

2. ordinary and partial differential equations
3. linear algebra, eigenvalues
4. probability and regression
5. transforms (especially Fourier)
6. symmetry (including group theory)
7. optimization (calculus of variations and linear programming).

The biologists in the US have a consortium type of organization, the Federation of American Societies for Experimental Biology (FASEB). At the present time, Mathematics Professor Maynard Thompson of Indiana University has been working on compiling the results of a survey done by the Service Course committee in cooperation with FASEB. The questionnaire, which has been distributed to biologists, biochemists, pathologists, nutritionists, physiologists, pharmacologists and immunologists through their individual societies, asks for information as to how much time students have within their program for the study of mathematics, statistics and computing. To determine where emphasis should be placed, part II of the survey contains a list of various topics (under headings of precalculus, calculus, differential and difference equations, linear algebra, probability and statistics and computing) which we have asked the respondent to rate as very important, somewhat important or unimportant. My own preliminary observations of the survey results reveal that the most serious mathematical needs were listed as first year calculus with series, some differential equations, with a heavy emphasis on applied statistics and probability along with a working knowledge of at least one higher level programming language, files and file manipulation, data acquisition and data reduction and computer graphics and simulation.. My own observations of the kinds of questions asked by some of my own students who major in biology lead me to believe that there is also a need for mathematical modeling, simulation techniques and optimization methods.We are still making contact with professional business schools and schools of health, nursing and medicine. In the US, architecture schools require little or no mathematics and do not seem to want to change; they rely heavily on teaching their own computer graphics techniques. We know from our own observations that social scientists need to know a fair amount of statistics, probability and modeling. The techniques in mathematical modeling in social science might be categorized as linear programming (game theory), recursive functions, matrix methods and eigenvalue problems at the lowest level. The course that I teach in mathematical modeling frequently attracts a few social science majors so I do lots of demography, arms race models, games, graph theory, harvesting resource models, Leslie models, Markov chains, linear programming and simulation. There is a lot of good mathematics in the analysis of these models; they require calculus, linear algebra, differential and/or difference equations, probability, computing and algorithm analysis, error analysis, statistics and graph theory.

The most pervasive comment we get from all of our contacts, both within our own discipline and outside, is that our students seem to pass our

courses but cannot DO anything. I would like to concentrate on that for
a moment for up to now I have discussed only *what* should be taught. But
a student is not a vessel; we cannot just pour it in. It is wrong to
think of knowledge as something we can transmit like a letter from
teacher to pupil. Piaget's investigations and more recent research in
cognitive science show that ideas are constructed out of action in an
appropriate experience. Are we providing such experiences for our
students? Are we guilty, as Sherman Stein (1985) says of entering into
a

> "secret pact between the teacher and the student: the
> teacher will not ask any question that will might
> reveal the student's ignorance, and the student, in
> turn, will not ask any question that may delay the
> class"?

Our need to learn more about cognitive science is acute. Some of the
results of recent research are useful and, yet somewhat discouraging.
Studies (Lochhead 1983) have shown that college students who had had a
semester or two of calculus still could not express relationships
between variables: Only 27% could answer this question correctly:

> Write an equation using the variables C and S to
> represent the following statement: "At Mindy's
> restaurant, for every four people who order
> cheesecake, there are 5 people who order strudel."
> Let C represent the number of cheesecakes and S
> represent the number of strudels ordered.

To show students the power and richness of mathematics we have a
responsibility to help them to gain conceptual mastery of the basic
ideas with a concommitant flexibility in the applications of these
ideas. In the face of their misconceptions, which have been shown to be
very difficult to change, we may be attempting the impossible. We
should be presenting problem-solving situations which require
resourcefulness and exploration. While there is always a place for
problems requiring the straight-forward use of the appropriate formula,
students should be taught a variety of techniques from which they can
choose. In all of this, the need for small classes, contact among
students and between faculty and students is an essential part. At most
universities in the US, Freshman Composition is a required English
course and sections are limited to about 15 students with personal
attention granted to each enrollee. We have been remiss in not
demanding the same personal contact with our students, for the ideas we
teach are no less important and our students might be far more ignorant
of our subject to begin with. To facilitate the attainment of our lofty
objectives, the best we can do is to help the student in:

1. learning to communicate mathematics
2. learning to do mathematics
3. learning to relate to mathematics
4. learning to learn mathematics.

My contention is that in the absence of substantial knowledge about
cognitive processes and the ways in which people learn mathematics,
those of us who have to teach beginning students must find a way to use
the little of what is known and our own good sense to improve
instruction.

1.*LEARNING TO COMMUNICATE*

Students communicate fluently in their native language when
they get to the university. A mathematician may therefore assume that
students can listen, read, write and speak in the mathematics classroom.
But we forget that to most students mathematics is a new language. The
language is a strange and difficult one. We use words differently in
mathematics than in general speech.   We use notation that is full of
meaning, expressible in whole sentences of real talk. We speak quickly,
we have accents, we write sloppily (was that a subscript or a multiplier
or an exponent?). We know how we feel we have truly understood a concept
after we explain it to others. Do we allow students that chance to
understand by active participation and communication?

The mathematics classroom is in many ways like a language classroom.
Skill in oral and written communication needs to be taught. Effective
listening is perhaps the most difficult. Students do not know how to
listen to what we say. Many of them do not realize the difference or
likenesses among
          if p, then q
          if q, then p
          q, if p
          p only if q
          p if and only if q.

The student needs to hear and understand a lot at once. We should take
the time to point out what is the hypothesis and what is the conclusion.
A little lesson in propositional calculus along with the theorem
statement goes a long way to clear things up. There is good reason to
spend some time on the predicate calculus. Students do not always know
what we mean when we say "for all ... there exists" or "there exists ...
for all". The $\varepsilon - \delta$ definition of the limit is a particularly sticky
one. But even worse than the use of quantifiers is the negation of a
statement with quantifiers. To a novice, it is incomprehensible to hear
"it is not true that for all ε there is a δ ....". How does a student
show that a set of vectors is not linearly independent? We need to
translate the negation of
          "non-zero vectors x_1, x_2,...,x_n are linearly

          independent vectors in a vector space V over $\mathbb{R}$ if and
          only if whenever
$$a_1x_1 + a_2x_2 + ... + a_nx_n = 0 \text{ for } a_1,a_2,...,a_n \in \mathbb{R}$$
$$\text{then } a_1 = a_2 = ... = a_n = 0".$$

They are rushing to take notes. They glance at the blackboard, they are
writing something the instructor said a few seconds before,
concentrating on what he or she is saying now. We can help our students
listen and participate in the active thinking process. We can speak
slowly and distinctly (no mumbling to the blackboard allowed). We can
use our voices to stress what is important - especially the little
words. And we can write legibly on the board exactly what we want the
students to be copying in their notebooks. What appears on the board
should be a logical development of an idea reasoned out with active
student participation. No erasing allowed unless there is an error. No
fair substituting x for w by erasing w in an equation they haven't even
got down yet. Rewrite it where it belongs in the logical sequence.
Pictures should be labelled and used often.

We should allow time for eye contact. We are particularly bad about
this because we always feel the need to get through the syllabus and a
quizzical look might shame us into backtracking. We must stop for
questions often - if no one asks one, be prepared to ask a few of our
own. Get students to summarize the lesson so far or to try a problem
using the newest concepts in an open ended way. Ask "is there a
solution?" instead of "find the solution". "What would happen if we
left out this part of the hypothesis? Can we find a counterexample?"
"Is the converse true?"

Mathematics and notation can be so confusing. One minute the
notation for a function is f. We are to treat f as a transformation.
Then we write $f(x)$, the value of the transformation at a particular
point. Then we take sums, products and roots of the transformations.
Or we operate on them (finding derivatives, for example) to get new
transformations. We need to express to students how we are using the
notation. Studies of student errors show the misconceptions. Being
aware of the pitfalls and the errors in thinking that cause them can
help the instructor to sound the warning. It is no wonder our
colleagues in physics and engineering complain that our students have
trouble dealing with *changes* in notation. Our students do not really
understand what a variable is! Research (Kaput 1985) into their errors
shows that students dealing with a Mindy's restaurant type problem
confuse the idea of a variable with the use of the letter to represent
units. They think of a set of cheesecakes consisting of 4C's for every
5S's in the set of strudels; the multiplier-coefficients become
adjectives in their language so they write 4C = 5S.

Listening is enhanced when there is motivation. We should present
interesting accessible examples **before** we offer concepts. A good
example will solidify the idea better than any definition. Problems
should be presented before the formal definitions and theorems. There
should be an opportunity for students to play with their ideas before we
formalize and prescribe solutions.

Students have to be helped to understand how to read mathematics. They
are trained to read quickly in many of their other subjects - they would

never finish the history or philosophy assignment if they were slow.
But being slow is just what we expect when we assign three pages of
mathematics. We figure it could take an hour to properly absorb the
material in such a few pages. Students don't know that; they barely
skim the material, hardly work the examples and never read it twice -
just go off to do the exercises at the end. Maybe our advice won't be
heeded, but it is worth a try to tell them that we expect three pages to
take them an hour's concentrated, active reading with pencil and paper
in use.

Students need to speak mathematics. I have banned the word "it" from
the students' vocabulary for at least a month at the beginning of a
course. It is amazing what a difference it makes. How do you read μ,
σ, ε, δ (could it be the script for the Hebrew letter, lamed?), $|x|$,

$f(x)$, $n!$, $\binom{n}{k}$? When students cannot read notation aloud, they don't read
it when they see it in the text, they don't identify it when you mention
it in class and pretty soon, they are quite lost and frustrated.

2. LEARNING TO DO MATHEMATICS (AND COMMUNICATE IT)
          I have used various forms of group work in elementary
courses. Typically, a calculus section at my university has about 25
students. I have taught, or rather the students have taught themselves,
a full syllabus of the course on several occasions using a small group
discovery method. This method, introduced to me by Neil Davidson and
Jerome Dancis at the University of Maryland, requires a great deal in
the way of preparation by the instructor. In a sort of Moore method
approach, the students are given examples to work out with guidance, a
form of programmed prodding toward a solution. When they have worked a
few examples, there is some motivation for a definition, the definition
is presented and the students formulate and prove the theorems. The
essential feature is that the class is divided into groups of 3 or 4 and
during the class period each group must work on the materials. The
instructor circulates among the groups answering questions, helping to
facilitate and clarify the learning. Students are talking, explaining
to others how they think they should proceed, there is arguing and there
are many blind alleys. BUT the students are doing mathematics, they are
listening to each other, they are speaking to each other and they are
active participants. Most important, the teacher sees and can guide the
students' thought processes. Here is the time for experimentation (try
a few numbers), geometric representation (draw a picture),
generalizations (would this work if...?) and summarizing. My preference
is that every few days each student hand in his or her own notes which I
correct and return promptly. They have nightly reading assignments and
problems to solve alone for homework and frequent quizzes and
examinations. Student evaluations at the end of the course invariably
state their satisfaction with themselves and what they have
accomplished. Some have said, "This is the first time that I felt I
understood mathematics".

Here is a unique opportunity to encourage women and minority students
who may be reluctant to contribute in class. I have found that the
women have to be told to speak up even in the small groups. At the
beginning they tend to let the men do the talking. The teacher can
direct questions at the reticent ones. They often have a great deal to
offer the group, and the rest of the group soon realizes that everyone
deserves their attention.

Students called upon to speak in class should be required to use
terminology and deductive reasoning consistent with the level of the
course. Although it is a painful experience at first, who would think
of teaching a foreign language without requiring practice in speaking
the language? When students realize the skills they are gaining in
learning to communicate their ideas, they feel very positive about their
classroom experience.

In studying applications and modeling even at the simplest level,
different interpretations of a problem give different answers. When are
two forms of a solution equivalent? If more open-ended problems were
assigned and students were encouraged to work together on a regular
basis, they would be forced to communicate.

Students need to write mathematics. Just as speaking forces them to
confront their difficulty - "I can't explain it but I know what I
mean."- writing in sentences to summarize a problem can prompt them to
consider what it is that they are doing. In service courses, frequent
assignments, promptly corrected, are essential. A great deal is
revealed in student papers. First of all, a particularly clever
solution is noticed by the instructor and the student reaps a little
well-deserved praise. Non-standard problems require explanatory
solutions. Communicating clarifies thinking.   Correcting assignments
also points up students' misconceptions. Answers are correct, but
reasoning faulty:

$$\frac{\sin 6x}{3x} = \frac{6 \sin x}{3x} = \frac{2 \sin x}{x} \quad \text{hence} \quad \lim_{x \to 0} \frac{\sin 6x}{3x} = 2.$$

Two wrongs can make a right:
Suppose that X has a normal distribution with $\mu = 30$ and $\sigma = 5$. Then

$$P(X \le 25) = P \left(Z \le \frac{30 - 25}{5} \right) = P (Z \le 1) = .5 - .3413 = .1587$$

Naturally, writing in proper sentences with correct grammar should be
expected.

Problem solving is very hard to teach. We can read Polya, we can study
Alan Schoenfeld's newest work on problem solving, (1985). We can
consult one or many of the new problem books. It can be extremely
difficult to accomplish a lot in a standard lecture classroom. We need

to assign and discuss multi-stage problems which may take a week or more to solve. We might, as mentioned above, encourage students to work in groups on both easy and challenging problems. We should assign problems which are best solved with a variety of techniques including methods making use of calculators and computers. We should ask questions which probe the conceptual bases- asking "Is it true that...?", "Find an example of ...", "Is it possible to ...?".

Perhaps paramount in helping students learn to think is the example the instructor presents in his or her own approach to problems. Neat, pat, and carefully prepared solutions to complicated problems are not as helpful as thinking out loud, occasionally showing a blind alley, analyzing by using problem solving strategy (draw a picture, try a few numbers,...), and bringing the enormous power of computer software into the classroom. Some of my colleagues argue that they have no time to use the available software with their elementary classes because the course outline is so demanding. It is obvious that all this communication requires time so that we may have to shorten the syllabi which have trapped both faculty and students with frustration. Surely it is better for students to have a solid understanding of the most basic concepts than to have a superficial acquaintance with lots of topics.

We would be remiss if we did not attempt to use symbolic manipulators in many service courses. Such algebraic systems, now available on microcomputers and hand calculators, allow for exploration and experimentation with complex problems. They represent the way non-mathematics majors (especially) will be doing mathematics in the future. Many of the factoring problems, the techniques of integration, complicated derivatives are quite beside the point for these students. They realize it too. Frequently, they are more conversant with computers than is their mathematics instructor. Those who have experimented report very satisfactory results. Research in this area continues at several US and Canadian colleges and universities.

3. LEARNING TO RELATE TO MATHEMATICS

There are many excellent resources available stressing the mathematics of the real world. Publications of the National Council of Teachers of Mathematics in the US, the new TEAM materials which include workbooks and videotapes (available from the MAA) and the wonderful COMAP modules (and newly produced videos) edited by Sol Garfunkel (see references) present realistic situations in which undergraduate mathematics is used. These are available for use in the classroom, or for extra assignments for the students themselves. They can be read by instructors and presented as examples even in large lecture classes. The role of mathematical models in the physical, biological, and social sciences can help students feel that the subject we teach is for them. Models in medicine, chemistry, economics, demography, voting and business can even make learning fun. Students reveal that they appreciate the time and concern that is shown by the instructor who makes the effort to present these applications.

Students can be given the responsibility of reading material using
mathematics in journals of their own disciplines or of preparing term
papers on their own ideas of how mathematics can be applied to another
field. Sharing these with the rest of the class gives a sense of
immediacy and relevance to the mathematics in the course.

4. *LEARNING TO LEARN MATHEMATICS*
          If we are able to help our students communicate, think and
relate concepts to realistic situations, we have gone a long way in
helping them learn how to learn. By making them self reliant, giving
them the confidence to try solving something they have never seen
before, we greatly enhance the possibility that they will be able to
learn new ideas.

We probably need to do some ambassadorial work as well. Students
frequently say that in their major courses (for which they were required
to take mathematics) instructors avoid using mathematics. This is
particularly true in the social and biological sciences in many schools.
It may be that the faculty themselves feel insecure with their own
mathematical skills, but we can be of help if we approach our
colleagues, ask for a consultation, and discuss what they want for their
students. This can be extraordinarily interesting. When I met with the
folks in our department of health science, I found that they wanted
topics in our courses that are not in the standard textbooks, while they
were perfectly happy to drop some other material which they felt was
outdated in the practice of their field. This personal relationship
between the departments can go a long way in showing the students that
mathematics is not an esoteric, irrelevant and incomprehensible
discipline.

At an advanced level, student teams working under faculty
supervision to solve industrial problems reap many benefits. At our
institution, the Applied Mathematics Laboratory accepts industrial
projects which are suitable for one year's work. Students and faculty
in mathematics are frequently teamed with those from other departments
to solve applied problems. Professional oral and written reports are
provided to the sponsor by the team. Faculty advisors learn about
applications of mathematics and they see opportunities for research in
mathematics. Other departments in the university and other industries
recognize that mathematics is applicable in a wide variety of
interesting situations.

Clearly, much research in mathematics education is needed. How can
teaching modalities change attitudes and achievement of the service
course student? Do students who are interested in science learn
differently from those who are in the social sciences or business? Can
we teach algorithmic (constructionist) thinking at the same time that we
teach the more standard mathematical thinking patterns? A recent
article by Don Knuth in the <u>American Mathematical Monthly</u> (Knuth 1985)

attempts to consider the similarities and differences in the two. Finally, the need for resources to allow for small classes, use of computer technology, and research in learning is severe. Conferences such as this one, however, can influence the improvement of the mathematical experience of the millions of students who sit before us in their obligation to their own academic interest.

REFERENCES

Albers, D. J, et al, editors (1985). New Directions in Two-Year
          College Mathematics. New York: Springer-Verlag

Committee on Discrete Mathematics in the First Two Years.
          (1986). Report. Washington, D.C.: Mathematical
          Association of America.

COMAP publishes modules, newsletters, video-tapes and the UMAP
          Journal, in applied elementary mathematics, COMAP,
          271 Lincoln Street, Suite NO. 4, Lexington, MA 02173.

Conference Board of the Mathematical Sciences (1981). Report of
          the Survey Committee, Vol VI, Undergraduate
          Mathematical Sciences in Universities, Four Year
          Colleges, and Two Year Colleges. Washington,
          D.C.: CBMS.

Kaput, J. (1985). Research in the learning of mathematics: some
          genuinely new directions. In New Directions in
          Two-Year College Mathematics, ed. D.J. Albers et al,
          pp 311-40. New York: Springer-Verlag.

Knuth, D. E. (1985). Algorithmic thinking and mathematical
          thinking. The American Mathematical Monthly, v.92,
          no. 3, pp 170-81.

Lochhead, J. (1983). The mathematical needs of students in the
          physical sciences. In The Future of College
          Mathematics, ed. A. Ralston & G.S.Young, pp.55-68.
          New York: Springer-Verlag.

Ralston, A. & Young,G. S. editors (1983). The Future of College
          Mathematics. New York: Springer-Verlag.

Schoenfeld, A. (1985) Mathematical Problem Solving. New York:
          Academic Press.

Stein, S. (1985). Routine problems, College Mathematics Journal,
          v16,n5, 383-5.

A FINAL STATEMENT

Mathematics is of increasing importance in all sciences and in everyday life. It is an essential part of the general culture needed by every citizen in order to understand our world and treat information and data with a critical mind. It is already an essential tool for many professions and will become necessary for many more in the future.

Mathematics has therefore to be taught to many students whom mathematicians have not considered before - to students of subjects as widely differentiated as home economics and biology. Even in the fields where a mathematical education is a tradition - such as physics and engineering - many changes are necessary. Advances in mathematical and computational tools make mechanical techniques and even skills less important than before. Mathematical understanding becomes even more crucial when students and professionals use computers, symbolic manipulation systems, computer graphics and other kinds of new technology. For the same reasons continuing education demands an increasingly important role. The successful design of mathematical courses to meet these needs requires an increased degree of understanding and cooperation between mathematics teachers and those in other disciplines.

All mathematicians must be aware that the future of mathematics as a science depends on the way they respond to these new needs coming from other disciplines and from society as a whole.

Public opinion and governments should be made aware of the urgency of meeting these new needs. The status of service teaching and service teachers must be improved. New appointments, new means and increased resources are vital.

<u>List of Participants</u>

Mathematics as a Service Subject

Udine, 6-10 April, 1987.

Isam Ahmed, Sudan
Samson Ale, Nigeria
Andrea Bacciotti, Italy
Reinhold Böhme, West Germany
Jean-Michel Bony, France
N. Boudriga, Tunisia
Dick Clements, U.K.
Leonede de Michele, Italy
José Ezratty, France
Hiroshi Fujita, Japan
Bernard Hodgson, Canada
Geoffrey Howson, U.K
Yoshihiko Ito, Japan
Jean-Pierre Kahane, France
Peter Kelly, U.K
Aderemi Kuku, Nigeria
Norbert Kusolitsch, Austria
Pierre Lauginie, France
Jack van Lint, The Netherlands

Emilio Lluis, Mexico
Bernard Malgrange, France
Angelo Marzollo, UNESCO
Eric Muller, Canada
Haruo Murakami, Japan
Tibor Nemetz, Hungary
John Newby, U.K
Mogens Niss, Denmark
Nicolas Patetta, Argentina
Henry Pollak, U.S.A
Elie Roubine, France
Tony Shannon, Australia
Martha Siegel, U.S.A
Fred Simons, The Netherlands
Dilip Sinha, India
Jan Smid, The Netherlands
Elisabeth de Turckheim, France
Vinicio Villani, Italy

SELECTED PAPERS ON THE TEACHING OF MATHEMATICS AS A SERVICE SUBJECT

Contents:

A.G. Howson et al   Mathematics as a service subject
                         (ICMI discussion document)
I.A. Ahmed   Teaching service mathematics: remarks
                         from a third-world perspective
A. Bacciotti and P. Boieri Teaching mathematics to engineers:
                         some remarks on the Italian case
R. Böhme   Teaching mathematics to engineers in
                         West Germany
P.L. Hennequin   La situation en France
B. Hodgson and E. Muller Mathematics service courses:
                         a Canadian perspective
A.O. Kuku   Mathematics as a service subject - the
                         African experience
T. Nemetz   A summary of the Hungarian educational
                         system with special attention to
                         mathematics teaching
J.C. Newby   The service teaching of mathematics in
                         Britain
N. Patetta   Mathematics as a service subject in
                         Argentina
A.G. Shannon and J.G. Sekhon Service mathematics in
                         Australian higher education
G. Aillaud   Mathématiques et discipline de service
G. Châtelet   Intuition géométrique - intuition
                         physique
G. Choquet   Some opinion papers given by members of
                         the French Academy of Sciences
J. Ezratty   An inquiry of the Association Bernard
                         Grégory
G.A. Jones   Teaching mathematics to mathematicians
                         and non-mathematicians
E.R. Muller and B.R. Hodgson The mathematics service course
                         environment
N. Patetta   Mathematics as a service subject - an
                         interactive approach
D.K. Sinha   The users' point of view
J. Tonnelat   Sur l'abus des modèles mathématiques

        This selection of papers, edited by R.R. Clements,
P. Lauginie and E. de Turckheim, will appear in the Springer-Verlag/
CISM series.